Askar Iskenderov   Nurserik Kudereev

The Canons of Evolution

The Canons of Evolution

Iskenderov
Kudereev N.

New Physics. Climate Threats. A New Theory of Everything.

In the book, The Canons of Evolution, the authors analyze the contradictions of modern scientific theories and, together with the reader, build an evolutionary theory of climate in the context of the general theory of evolution, as well as the hypothesis of the Solar System formation – the emergence of the world from the chaos of the vortex first beginning – based on new physics postulates. One of the book's main tasks is to understand the true cause of climate change and present the authors' vision of the theory of everything to the world.

Editor A. Meshcheryakova

TABLE OF CONTENTS

Aristotle and Windmills

Chapter 1. The Beginnings of Physics and the Foundations of the Theory of Evolution

Fundamental Physics: Questions Unanswered
Physics and Philosophy. Postulate Foundations
The Choice of the Substrate of Evolutionary Movement
Two Forces of Newton and the Primordial properties of Amions
Medium as a Physics Category. The Concept of Ether
Vortex Formation
Evolutionary Medium

Chapter 2. The Formation of the Solar System

First stage of the physical evolution of the world
Second stage of the physical evolution of the world
Third stage of the physical evolution of the world
Fourth stage of the physical evolution of the world
Fifth stage of the physical evolution of the world

Chapter 3. Evolutionary Theory of Climate

What is climate? Some clarifications
Winter House: The Climate Model on the Window Glass
Two Levels of Climate
The Physical Essence of the Climate
The Formation of Near Space
The Formation of Crust. The Formation of Atoms and Matter
Implosion: The Stick Effect and Quantum Mechanics
Variability of the Medium is the Cradle of Evolution
Annual Climate Cyclicity
Biological Life's Exit to Land
The Magic Properties of Water. The Water Cycle in Nature
Coincidences in Nature is the Law of Evolution. Climate Stabilization

Biocenosis. Acceleration
The Gentle Level of Climate
Epochal Cyclicity of the Climate
Average Temperature of the Planet and the Climate Entropy
The Main Causes of Climate Destabilization
Vortices and Other Phenomena in the Lower Atmosphere
Preliminary Results: 7 basic theses of the Evolutionary Theory of Climate

Conclusion

Bibliography

Figures

> These two acts are but one,
>
> under different denominations.
>
> The unique act of generation; that is the mystery of the beginning;
>
> the mystery of mysteries;
>
> the door through which have issued, onto the scene of the universe, all of the marvels which it contains.
>
> Lao Tzu

Aristotle and Windmills

In my childhood, I really enjoyed building models and various technical things, dreaming that one day I would become an inventor. Once, already as an experienced engineer and participant in a republican contest, I won a grant for the development of a wind farm.

It is important to say that in search of engineering solutions, I always try to rely on theory, so I immersed myself in the scientific part of the technical task. Doing what I loved, I began to collect information about everything related to wind power, exercising my mind with 'complex sciences' when I had free time. I was also interested in related industries and applied physics in general. As usual, simple questions lead to a dead-end road.

From our school program, we know that the heated air goes up and gets replaced with the cold wind. In the real world, heavily heated air can stand still for months; this is called an anticyclone. When it disappears or moves, it is impossible for the weather service to predict weather accurately. The reason bothered me.

All inventors of wind turbines (windmills) have one concern, namely, waiting for the wind; and there is one trouble, the stormwind. Suppose the first is an anticyclone, and what about the second? I was amazed to learn that I was fighting like Don Quixote, only in my case, there were fragments of climatic vortices reaching a few thousand kilometers in diameter. Scientists call them cyclones.

Studying the nature of the wind and delving into climatology step by step, I found out a strange thing – science does not explain the nature of even ordinary steppe vortices. They come suddenly, encouraged by the internal forces; however, neither physics nor other 'complex' sciences have an answer as to what these forces are.

Climate change has yet to cause me much anxiety; I treated it like all ordinary people, considering it to be a long process that exists by itself. But I have always wondered - how does a brutal whirlwind or tornado raise the whole house high in the sky and gently carry it along with the inhabitants?

At first glance, the force of the wind should destroy things. However, the vortexes can, deftly manipulating their invisible 'hands', carry away and put back exorbitant weight. Trying to solve this trick, I could not find an unambiguous and reliable interpretation.

According to physics, the high efficiency of natural vortexes is explained by the force of vacuum, and the vacuum is a void. It is not clear how the force, the efficiency coefficient, which is greater than one, appears from the void. Trying to achieve the maximum efficiency, I decided to design an unusual model of a windmill and create an artificial vortex; I planned to use the lifting power of the mysterious vacuum. ("The universe is a large-scale vacuum fluctuation." Edward Tryon).

Having achieved noticeable progress in designing – although only due to the usual effect of cavitation – I recognized the need to study different types of forces and master the **concept of energy**. Working with the wind energy, I could not understand its essence. Besides, I had a hunch that in the period of the anticyclone, there is an accumulation of some excess energy, which is then dissipated in the form of cyclone vortexes. Later, this hunch turned out to be decisive in the creation of the theory of climate.

The idea required elaboration, and it was necessary to start from the very beginning, with some kind of a conditional point. That of Euclid or Aristotle? Does not matter.

Newton's mechanics calls the absence of motion, and hence of energy, a state of inertial rest. Looking around, I did not see anything that would not move. **Space** is the only thing in the universe that might not move, but physics has already crushed this dogma, and it is unclear why.

I imagined energy in the form of a substance (I was taught this way), which is separate from matter. It looked like this – a material particle and an antiparticle meet, destroy each other, and disappearing, turn into radiant energy. At the same time, physics argues that the light ray is the movement of the smallest photon particles; that is, it is still matter. This meant that matter does not disappear; it only reaches the maximum speed.

Thermophysics is a statistical mechanics; it regards air molecules as carriers of energy (heat). Why not photons, I asked myself, which deliver the bulk of energy in the form of solar radiation to the planet? In the search for answers to many 'whys,' I was convinced that there were significant discrepancies in the fundamental fields of knowledge.

Traditionally, it is believed that physics is a developed science, but going deep into it, I found something totally fantastic! Each of its sections builds theories on separate independent postulates, which often differ from the postulates of other sections and even contradict them. This was a complete surprise for a person accustomed to technical regulations and theories like myself.

Scientists invent theories. Engineers embody them in various devices. Inventors often come up with something contrary to the existing theories. Sometimes ordinary grease monkeys create such interesting things that advanced science has to catch up with them!

I can also be described as an engineer and as an inventor on the path of creative research; this is my special expertise and secret weapons. Yet I was very puzzled – what exactly am I missing? Or was something wrong with science itself? The help appeared in time.

A bright representative of the 'golden age' of Soviet physics, I.M. Khalatnikov, shared his views on the main scientific problems. In a TV interview, he put it briefly and simply, saying that **"the world science does not have fundamental physics."** Such support from an academician of the Russian Academy of Sciences and a foreign member of the Royal Society of London, a theoretical physicist, one of the leaders of the post-war atomic project, the founder of the Institute of Theoretical Physics in the USSR, inspired me to study the problems he outlined. I got bold and began to purposefully master modern scientific theories.

The windmills, the wind, the vortexes, nature, and physics rolled into one ball in my head. I can put it another way – the combination of circumstances dragged me into the 'scientific morass,' like a non-Newtonian fluid. During this time, picking up the pieces of impartial information, I became a convinced ecologist and literally began to feel the effects of climate change on my skin.

As I immersed myself in the actual problems of different sciences, an analyst awakened in me. Aristotle wrote in his Metaphysics, "The craftsman knows 'what' and the mentor knows 'why', so we honor the mentors more," and "The first beginnings and reasons are the worthiest of knowing because everything else is known through them and on their basis, rather than being known through what is subordinated to them."

It all started with indivisible particles and self-education; I started everything from scratch, testing all the concepts of fundamental physics.

I was lucky that I was not a physicist and did not have the pressure of **old concepts and terms of science (or its clichés)** on me.

The years of searching turned into quality – a new world outlook, or a new system of views on the world.

Chapter 1. THE BEGINNINGS OF PHYSICS AND THE FOUNDATIONS OF THE THEORY OF EVOLUTION

Fundamental Physics: Questions Unanswered

Modern knowledge has many gaps in the form of unproven or incomplete theories. The origin of the world and the appearance of life on the planet are still compelling mysteries. In my opinion, the lack of a consistent hypothesis about the beginning of the world is a flaw of fundamental physics, the task of which is to determine the driving forces of the first evolutionary processes, the laws of interaction of these forces, the basic properties of matter-primary element, and its other characteristics.

As a rule, scientists only consider the force of gravity as the initiating force in numerous theoretical constructions. The main drawback of this approach is that it is unclear what exactly causes cosmic dust and light gases to vortex. For example, the essence of the **hypothesis about the vortex origin of the world according to Kant-Laplace** is that a cosmological substrate consisting of particles is concentrated and compressed by gravitational forces to the state of plasma. Then, the external forces, or cosmic randomness, should 'ignite' or launch the processes of thermonuclear reaction and 'light' the Sun.

As a generally accepted version for the general public, I will give a brief note from Wikipedia. "According to the hypothesis of Kant-Laplace, the site of the solar system previously housed a huge gas-dust nebula (similar to the smoke from the chimneys – **Author's note**). According to I. Kant, the dust nebula consisted of solid particles; according to P. S. Laplace, it was gaseous. The nebula was hot and spinning. Under the influence of the laws of gravitation, its matter gradually thickened and flattened, forming a nucleus in the center. This is how the primary sun was formed. Further cooling and compaction of the nebula led to an increase in the angular velocity of rotation, as a result of which the outer part of the nebula on the equator separated from the main mass in the form of rings rotating in the equatorial plane; there were several of them. Laplace cited the rings of Saturn as an example. Unevenly cooling, the rings burst, and as a result of attraction between the particles, planets orbiting the Sun were formed. The cooling planets were covered with a hard crust, on the surface of which geological processes began to develop."

In this version of the hypothesis, I do not find an answer to the question of why the primary gas-dust nebula was rotating. Why did it "condense under the laws of gravity" when the gases cannot be compressed to the state of plasma with the help of very weak gravitational forces? Why did the planets start to rotate around their axis? Finally, I believe that arbitrary self-cooling in space is an incorrect interpretation of the process; this approach is totally incorrect. This issue is directly related to the climate and will be considered in detail in the relevant chapter of the book.

It is also unclear how the planets were formed separately and how they separated from the protostar, a large vortex. Some scientists suggest that the satellites of the planets were formed during the collision and destruction of the planets themselves. When proclaiming the ideas of determinism or continuous evolution, Laplace admits that there are gaps in his evolutionary hypothesis, including the initial half-preparedness of the basic world in it. If he does not find a rational objective reason, he explains any stage by cosmic chance.

Physics as a whole does not take into account the basic role of the primary matter in the arrangement of the world. Cosmic dust consists of small particles of various minerals. Why is it taken in the form of a substrate and not the primary matter?

There is no theory of atoms' 'self-assembly' in physics. Kant, Laplace, and modern evolutionists take 'raw materials' from the finished atoms as a substratum of evolution in their hypotheses of the solar system formation. They do not take into account that planets, their satellites, and chemical elements are born in the process of solar system evolution. Then comes the biological life, followed by the humans. And it all begins in the galactic space from the primary matter.

Physicists adopted cosmic dust and gases consisting of chemical elements (atoms) as a substrate of the original cosmic nebula. It appears that they begin to consider the evolution of the solar system from the middle point, or some intermediate point of reference. The question, who prepared the space's 'semi-finished products' for launching the creation of the world and how they did it remains unanswered.

Kant and Laplace's ideas about the origin of the world appeared after the discoveries of Galileo, Kepler, Descartes, and Newton. Galileo discovered the inertia of objects. Kepler discovered orbital laws based on Tycho Brahe's data. Descartes found out the vortex nature of the world. Based on all of this, Isaac Newton composed the foundations of mechanics, into which he introduced physical concepts of inertia, force, acceleration, and mass.

What did Newton do for science in this direction and what did he not clarify? By creating the mechanics of the planetary system, he included the forces of gravity in it; they attract the planet to the Sun and continuously reject the planet's inertial impulse to rectilinear motion. Two properties of cosmic bodies have been identified: their inertia and gravitational forces. At the same time, he also included the 'divine impulses' in his celestial mechanics to help the two properties.

Later, Laplace created a mathematical model of the solar system and proved that the planets can move in a circular (elliptical) motion by inertia for a long time – or forever, at the limit. He came to the conclusion that Newton's 'divine impulses' were not needed at all. It was as if Laplace finally removed – or solved – the veiled **fundamental question of physics and philosophy** and the eternal dispute between materialism and idealism. However, Newton's 'divine impulses' and his 'unscientific' ideas have yet to become clear.

Interim conclusion. *A close look at modern hypotheses of the world evolution gives rise to many questions. Why did the planets separate from the large vortex and the total mass of stellar material? Who 'ignited' the thermonuclear reaction in the star's proto-body from the outside? How did the heavy chemical elements get into the body of the planets? In the Chapter, "The Formation of the Solar System," we will try to answer these questions, developing a consistent hypothesis on the basis of new physics principles and postulates accepted further in the present chapter.*

To further specify my questions and doubts, I want to turn to Richard Feynman's *The Character of Physical Law*. Feynman is a well-known physicist, Nobel laureate, and participant in the US atomic project. In his book, he explains very difficult questions in accessible language using simple examples and the right words. At the same time, Feynman is a typical representative of modern physics, the bearer of its ambiguous ideas.

He writes: "Free movement has no apparent cause. We do not know why objects are able to fly forever in a straight line. The origin of the law of inertia remains a mystery." [60]

It is understood that the free movement of an object is a movement by inertia; in particular, Feynman analyzes the forces of gravity. Let us look at the 'heavenly powers', for which we will consider the issue in its graphic representation. Figure 1 of this book is a copy of the figure from Feynman's book. It shows a planet orbiting, a vector of gravity forces directed toward the center, and a vector of the planet's inertia tangential to the orbital path.

Feynman's explanation of this figure: "Therefore, Newton decided, the planet orbiting the Sun does not need force to move forward; if there was no force, the planet would be flying on a tangent" [60]. Trying to simplify the presentation of important issues for the general public (or considering them already resolved), he distorts the history of science and says nothing about the fact that Newton supplemented the inertial motion of the planets with 'divine impulses'. Perhaps, Feynman believed that Newton was mistaken, and modestly kept silent about it.

My thought is that Newton is just being cautious. He understood that if the forces of gravity had caused an accelerated deviation of the planets towards the Sun, then, sooner or later, the planet should have fallen on the Sun, so he was forced to admit an occasional outside intervention. There must be some forces that push the planets and return them to the main orbit.

As the greatest physicist, Newton saw an inaccuracy in the planetary balance of forces, so he added 'divine impulses' in his notion of the planets' inertial motion. Followers later distorted his divinely dialectical approaches in physics, trying to interpret the laws of mechanics in the light of materialistic ideas.

Feynman fully supports Laplace and also believes that external influences are not needed. As a mathematician, Laplace ignored this physical 'wrongness' of Newton and completely mathematized celestial mechanics. It can be said that he hid the unsolved question of physics in the deep wells of faceless mathematical formulas. Later, James Clerk Maxwell proposed his integrated circuit and 'invisible gears,' trying to explain the electromagnetic field.

I completely share Einstein's opinion that all of them provided only random mathematical algorithms. The curtain of formulas concealed the physical, or philosophical, essence of the phenomena that they describe. As a result, theoretical physics lost rather than gained, having lost its metaphysical component and the stimulus that moved it throughout the ages. However, I believe that intuition can be considered one of the effective methods of science. For example, Maxwell's mathematical algorithms have been tested by practice and are currently in demand, while his 'invisible gears' still remain incomprehensible to us.

I believe that the distortions in the examples I cited, like many others, are connected with the interpretation of works of the great scientists of the past and with good intentions to facilitate and simplify the learning of the material by the students. New generations, separated from the original sources and studying the proposed interpretations, are deprived of natural scientific stimuli. Feynman avoids the 'divine impulses' of Newton, while other interpreters remove Mendeleev's 'ether' from the periodic table. The error path in science is paved with good intentions. Both Newton's 'divine impulses' and Mendeleev's primary element will participate in our further theoretical constructs in the most active way.

At the same time, Feynman writes: "Physics has not yet become a single structure, where each part is in its place. For now, we have many details that are difficult to fit together [60]." He notes the following as an example in his other work: "Until now, no one has been able to imagine gravity and electricity as two different manifestations of the same essence" [60]. In this book, we will take into account his conscientious recognition of the 'imperfections' of physics and try to find ways out of this situation.

Let us take a closer look at Figure 1. It looks much like a parallelogram of the balance of forces. In the figure, we see a dynamic equilibrium. However, it has nothing to do with the balance of the two forces. The vector of the planet's inertial motion can be described as a product **mv**, and the vector of the forces of attraction is **ma**, where **m** is the body mass, **v** is speed, and **a** is acceleration. There is a paradox; we made a parallelogram of equilibrium between the amount of motion **mv** and force **ma**!

I consider this an example of unintentional sophistication in the history of science. We compared (added) the surface area of the body with its volume, made a dynamic equilibrium of the planetary system, and faced the following dilemma: to fulfill the equilibrium condition and observe the rules of mechanics, we must either assume that the forces of attraction are not a force, or consider the amount of motion **mv** – inertia – a force.

In any case, the problem has to be solved, and we will do this later. The equilibrium of the planetary system and the laws of celestial mechanics exist in practice; this is an indisputable fact. The error probably lies in the physical concepts – most likely, inertia is not really inertia. We once again see the unresolved problems of fundamental physics. Does this mean that Newton's mechanics is wrong? If so, why did the future constellation of famous physicists miss the mistake?

The fact is that the forces of gravity arising between physical bodies behave like **ma**, i.e. they have a quadratic dependence – an inverse relationship to the distance between the bodies. Gravitational forces act at a distance, or in a non-contact way.

On the contrary, dynamic force **F=ma**, adopted (invented) by Newton, does not depend on any conditions and acts only with direct contact of the bodies. Let me clarify. Dynamic forces do not depend on the distance between bodies **because** they act during direct contact. Thus, we come to the assumption that the forces of gravity are something different in nature.

But let us get back to Feynman's book, where he writes: "This amazing test showed that Newton's theory is fine" [60]. Feynman comes to this conclusion when describing the set of experiments where the value of free fall acceleration **g** is determined.

I emphasize that Newton's laws operate between physical bodies. The question that arises is: how does this interaction occur in **micromechanics**, which, today, is called quantum mechanics? It turns out that there are irreconcilable contradictions between it and classical mechanics. Given the objective difficulties in the reliable knowledge of the microworld (the world of elementary particles), maybe we should check and clarify the mechanics of Newton?

When I shared my doubts with physicists, many jumped to their feet and began to write formulas on the blackboard, urging me not to "profane" Newton. They told me that Newton's laws are proven empirically. I will tell you how my doubts were resolved in the chapter, "The Formation of the Solar System." We will try to understand all the issues I mentioned on a great cosmic example.

Now let's touch on the poorly developed topic of vortex formation. What are the logical gaps in Laplace's hypothesis and his determinism? He does not explain what forces in space or, perhaps, the immanent properties of matter propelled the cosmic formation from particles into a circular and eternal vortex motion. It is clear that the forces of gravity can concentrate the particles, but it is unclear what forces got a large space vortex going. This 'thoughtlessness' in science has been going on for a very long time – for hundreds of years. We recognize various physical laws and explain existing phenomena, but we do not know what forces initiate ordinary natural vortexes in the atmosphere and the aquatic medium. In physics, there is still no theory of ordinary vortexes.

In the general system, a rotational force moment appeared as if from nowhere. We can only guess whether every speck of dust knows its circular route or the vortex movement started as a result of a gravitational cluster of particles. The only thing that is clear is: scientists have noticed that sometimes there are vortex effects in the atmosphere or the water medium. Hence, according to the principle of analogies, a general conclusion is drawn, which is that the cosmic accumulation of particles has or is initially capable of vortex motion. As is known, the analogy is a method that does not always have a large evidentiary weight; it must be combined with the general composition of science and the postulate base.

The forces initiating the vortexes could not just appear out of nowhere. And we see that the question of the physical meaning of the vortexes formation in nature, as well as of the forces that initiate them, still remains unanswered. What can I say: for the modern science, the nature of energy in general is unclear! The concept of energy in the literal and figurative sense exists by itself; it remains beyond the limits of the knowledge we possess. One might say, I came to this conclusion by chance, studying the climate and trying to comprehend what heat is in the atmosphere.

Summarizing all of the above, I would call the hypothesis of constructing an atom, the concept of matter, the concept of energy, and the theory of fields the main unresolved problems of theoretical physics. Going deep into the study of all these problems and issues, I made an unexpected conclusion that there is still no rational solution and answer to them.

On the level of the world outlook, the beginnings of fundamental physics have not progressed one iota since Newton times.

Before that, like many of you, I considered physics to be an unattainable science for myself, and physicists themselves the celestials. In reality, it is a colossus on clay feet; a mixture of mathematics, fragmentary facts, and physical concepts not connected by a general theory. Different sections of physics have independent postulates; sometimes they categorically contradict each other. There is the greatest confusion in the fundamental part of physics, and the main trouble is that we are not yet aware of how much we are mistaken.

I think that the main reason for all the shortcomings is that no one sets a time limit for solving these scientific problems and takes purposeful measures for the progress in fundamental physics backed by adequate resources. Applied sciences are successfully developing, and the main fundamental issues remain unanswered for tens (or hundreds) of years and are not even on the agenda.

For many reasons, experimental physics cannot help in this; this is the case when a scientific theory is needed first so that it becomes possible to prove it on the basis of at least indirect experiments. For now, we can only fantasize and resort to metaphysics, the cognitive method of Aristotle. In addition to metaphysics, I will also use the scientific method of Sigmund Freud.

Reading Freud, I admired his popular and, at the same time, quite boring lectures. He created the basics of psychoanalysis literally from scratch; slips of the tongue, forgetfulness, dreams, and all other trifles served as scientific material for him. He would scrupulously collect and accumulate minor indirect facts, separating the essential from them; as a result, he came up with the theory of psychoanalysis.

In this book, we will try to reproduce the main provisions of physics, adding new concepts for finalizing the hypothesis of the beginning of the world and the creation of an evolutionary theory of climate with a view to develop – sooner or later – the **general evolution theory**.

All of the above questions were raised in my mind as I became more aware of the climate threat. Over time, they merged into one big question: **How do we resolve the existing contradictions and fill in the gaps in fundamental physics?** In *The Canons of Evolution*, I want to, first of all, pay attention to these gaps and, secondly, try to fill them myself. But let's not get ahead of ourselves.

Physics and Philosophy. Postulate Foundations

Most scientists around the world, not only philosophers, highly appreciate the laws of dialectics and consider them the universal laws of world development. In particular, the famous **principle of unity and struggle of opposites** points to a universal source of development on a fundamental level. This principle, or law, of dialectics states the following:

1. *All objects, phenomena, and processes of the material and spiritual sphere necessarily contain contradictory principles and have contradictory sides and tendencies.*

2. *The struggle of the conflicting sides of an object occurs because the opposites are in unity; this prevents them from being destroyed; even more so, they improve and evolve.*

The unity and struggle of opposing principles are inherent in any physical body and material particle. The effect of this philosophical law should extend to the microcosm and determine the properties of the substrate, or the smallest indivisible particles. If we adopt a dialectical concept and use it critically in the construction of a general theory of evolution, then we will come to an unambiguous understanding of the initial stage of the evolutionary processes in the solar system.

However, before we propose our own hypotheses and theories and try to answer the fundamental questions, including of the natural sciences, it is necessary to develop the new beginnings of physics on the solid foundation of the postulate. We will take the paths of Aristotle and Freud – using both scientific approaches, we will consistently introduce all the necessary postulates and concepts that will later help in the development of hypotheses and the creation of theories. In *The Canons of Evolution*, this will be **the hypothesis of the solar system formation** and **the evolutionary theory of climate (or ETC)**.

We will have to rely more on circumstantial facts, which have one big drawback; namely, they are often interpreted ambiguously. There are always many options when interpreting indirect facts; their number is limited only by the capabilities of the human brain. Therefore, to build our own adequate and consistent interpretation, we will use well-known scientific, including philosophical, methods. I have already singled out dialectics as the main method.

Less common approaches to scientific creativity may also come in handy. I have already mentioned intuition and educated guesses; now I will dwell separately on another two. They are **the principles of hermetism**, of which the principle of correspondence will be the most important for future constructions: "what is above, is also below; what is below, is also above." To check the selected interpretation, we will apply a method that I call **compositional,** where the truth is proved by a system of concepts or as part of an acting model.

In school textbooks and various educational guides, problems are often given with the solution at the end of the book; students consolidate the material covered, and they can always check the correctness of the solution themselves. Offering these new theories to the reader, I do the opposite; we already know the future solutions to the problems. The reader should consider the ready solution and see that if everything goes smoothly, then we are on the right track.

Checking our hypothesis of the self-development of evolutionary processes, the solution of the problem is known in advance – it is the existing world. We need to create it theoretically. In my hypothesis of the formation of the solar system, I took the theoretical path, from chaos to our days without serious remarks. The evolutionary theory of climate was built on the same principle.

Consider the following situation. Let's say a device is broken. The master will first conduct a general inspection, make a diagnosis, and then begin to check the performance of each individual component until he detects the cause of the failure and replace the faulty parts. The main test of his work will be the successful launch and further proper functioning of the entire device. The fact that it starts and works means that the master correctly performed his work.

In our case, to create a hypothesis about the beginning of the world, it will be necessary to first determine the substrate and its properties necessary for self-launching of the evolutionary system. Considering the evolutionary processes taking place in the future, we will need to describe the conditions required for their appearance and development, general characteristics, and basic principles of functioning. If our scientific model, or evolutionary system, is successfully 'launched,' and it should be launched arbitrarily, without external influence, this will mean that the postulate adopted by us is correct. I emphasize that our system should start self-arbitrarily, without "accidental impulses" – the physics 'safe' should open without cracking.

Why is it necessary to start our theoretical constructions from the beginning of time? Because, and this will be our **first thesis**, there was chaos on the site of the galactic space, where the solar system was formed. Chaos is an antithesis to the order; it is not subject to any laws, including physical ones. In the world of chaos, only the primordial properties of substrates existed, and there were no laws at all.

The Choice of the Substrate of Evolutionary Movement

2500 years ago, the ancient philosophers tried to find the driving force of the world, the 'energy of life.' The concept of the substrate as the smallest and indivisible material particle in metaphysics was introduced by Aristotle, who believed that the substrate must have two opposite properties and emphasized its dual nature. At the same time, he clarified this in his book, *Metaphysics*: "Double does not mean two [2]."

In those days, people already suspected the existence of the primary element. The ancient philosophers had no tools for the study of nature and the creation of metaphysics except the power of the mind and fantasy. They were limited only to external observations of nature, but they managed to look inside, into the essence of matter.

How slowly the beginnings of science develop! Two millennia later, we are once again forced to look for the substrate of the world, search for the universal particles and their properties, and get back to our initial positions. Strangely enough, armed with perfect tools and technologies, today, we, like Aristotle, must resort to the help of metaphysics and the power of the mind when defining the substrate and its properties.

In physics, there are too many substrates – several dozens of them. In fact, all the research of the physicists in the 20th century was aimed at finding the substrate, or the 'divine particle.' Taking the decision to specify the substrate, I proceed from the fact that the initial chaotic accumulation of the pre-atomic matter in space did not consist of those elementary particles that are known to physics today. In a chaotic cluster, particles and antiparticles, electrons and protons (quarks) cannot be together randomly– they will destroy each other. I mean, the theoretical side of the matter with a hint of the modern hypotheses' flawlessness.

To build a new hypothesis of the formation of the Solar System, we need a substrate of the pre-atomic matter. Obviously, we will face great difficulties: the microcosm of elementary particles. Like the human brain for Freud and psychologists in general, it is difficult to access for empirical research. Therefore, we will have to arbitrarily assign the substrate and select its primordial properties, or **a priori properties**, from those in nature familiar to us.

When determining the a priori properties of the substrate, we will have to use indirect empirical information from the macrocosm, as well as directly use analytical methods; in particular, the **analogue method**. In science, this is called metaphysics, or direct philosophizing.

Dust and light gases do not stand up to criticism as a substrate of evolution – they cannot self-compact to the state of plasma, as gravitational forces are too weak for this. Let's ask ourselves whether such forces even exist in nature. It turns out that we already know them. They are the nuclear forces of interaction that arise between elementary particles. However, we must take into account that we do not reliably know the nature of the atom and how the atoms themselves appeared. Only by answering these questions can one know the true nature of the nuclear forces.

Therefore, to be able to move further in this direction, I introduce a universal particle in the form of the substrate of evolution, namely, some abstract **amion**. The new name was coined in order not to be confused in the terms and special names of elementary (pre-atomic) particles.

Amions are the smallest indivisible particles that have a priori (original) properties. The entire world consists of amions. This postulate, adopted through assumptions and guesses, will become our encryption key for the legitimate 'opening' of the secrets of physics.

Unfortunately, none of the physicists considers the place and role of tiny particles (pre-atomic matter) in our universe. At the same time, scientists cannot deny the fact that, objectively and continuously, there is a pre-atomic matter in the atmosphere and space. Directly or indirectly, weakly or thoroughly, it should influence the state of our world and the planet's climate. In physics, there is recognition that the substances (atoms) on the Earth's surface produce radiation, and we also know about the dual nature of the rays. These two facts mean that all substances produce emission of waves and emanation of particles into the atmosphere.

It is on the basis of these ideas in this book and, of course, in the general theory of evolution, that a new theory of the universe and climate theory will be built. In our hypothesis of the formation of the solar system, the world will consist of amions as a substrate or pre-atomic matter. In general, the book is built on a common – evolutionary – idea of its self-complication. From the very beginning, we will consider the role and place of primary matter in the evolution of the physical and then biological world, and trace its complexity from the initial chaos of zero time to our days.

Amion, as a new scientific and philosophical category, differs from the old transitional concepts in that it is both a substrate and a demiurge in one substance, in one tiny particle. In the history of science, including philosophy, the atoms of Democritus, the water of Thales, and Ostwald's energy were the substrate as the initial cause of the macrocosm. Today, we can call elementary particles, plasma, and energy as a substrate. Aristotle's approach was the most successful: he introduced the concepts of matter and form into philosophy. In this context, amion is the primary matter.

The demiurge is the creator of the visible cosmos. This concept was first used in ancient times before the formation of world religions. The term demiurge in this interpretation was introduced in the philosophical lexicon by Plato.

Amions are an integral part of complex material structures; at the same time, they themselves are able to compose complex structures. Amion is ***a detail of the evolutionary construction kit for any future product, a life-giving brick of the universe,*** as well as an updated philosophical concept of *monad*

The main difference between an amion and a monad is the concretization of its simple physical properties. The monad is known mainly in the last version of its interpretation as presented by Leibniz. He represents the monad as an indivisible unit of the universe and a self-sufficient substance; at the same time, there are many monads in the world; everything has its own monad. For Leibniz, it is an inimitable and spiritualized particle. In the postulate of the new beginnings of physics, which we are now forming, amions are universal particles. All kinds of substances consist of identical primary elements – amions – which have opposite but compatible properties.

I want to emphasize the similarity of the position of ancient thinkers and modern physicists: they are trying to create a world of two primary elements (substrates) with opposite properties. In turn, we will only consider one primary element, but with two opposite properties. This is the key difference in the proposed postulate; we follow the principle of dialectics.

Why is it impossible to empirically detect amions separately or piecewise? Because they are the corpuscular part of electromagnetic waves, as well as the carriers of gravitational forces. Can we explore water molecules with the help of sea waves? Of course not. Sea currents and waves can also react to large objects, such as an iceberg, defining (showing us) its underwater and bigger part with the large bend of the current.

By analogy, with the help of electromagnetic waves, we see (detect) only large objects inside the atom; we cannot see individual amions. Generally speaking, if we look into an atom to study it with the help of electromagnetic waves, does this mean that we are disrupting the balance inside the atom? Tens of years of scientific work and huge financial resources invested in the LHC (the large hadronic collider) have not helped us to find the substrate – 'the God Particle' – in an empirical way. This is because the smallest indivisible particles (amions from our postulate) are indeterminate and inaccessible to the empirical method of study; that is, perception through experiments.

Therefore, at the postulate level, we assume that the atom consists entirely of amions, which also form its medium, which is in structural equilibrium and is non-discrete (see the section Medium as a Physics Category). We will refer to the medium consisting of amions as an **amionosphere**. It has the ability of self-complication – vortex formation – under certain conditions due to the amions' two a priori properties, which we will discuss in the next section.

The postulate of any science is often a kind of tautology. Adopting the **second postulate** of the new physics beginnings (to recall, the first and the initial one is the postulate of the chaos of the first beginning), we can say that the substrate is the amions. I might as well as say that the amions are the substrate. The point is that all elementary particles known to physics (electrons, protons, and so on) will be considered as consisting of amions. This is the difference between the concept of the amion as the smallest indivisible particle, or pre-atomic matter, and all the elementary particles already defined by modern science.

Thus, forming new postulates of physics, we will consider amions as the substrate. Now, to move on, we must voluntarily assign them a priori properties. It should be noted that we will observe the rules of symmetry, which Feynman mentioned in his *The Character of Physical Law*, dedicating an entire chapter – 'The symmetry of the laws of physics' – to it.

It seems to me that modern physics has certain gaps in quantum theory, primarily because it needed to initially speak in terms of the primary element's properties (inertia and attraction) rather than forces. Therefore, we will introduce new concepts different from the physical concepts introduced by Newton into physics and, accordingly, into micromechanics. Previously, we had the following dialectical pair in macromechanics: **dynamic forces – forces of gravity**. Now we will form a similar pair for amions in micromechanics: **inertia – attraction**. I will refer to the attraction as **aser**.

In The Canons of Evolution, there will be two new concepts named using Kazakh words; they are aser and sulde, which will be introduced into the theory of climate in the third chapter. The book is written in Russian, so any new concept named using a word from a given language will refer us to this word's etymology and its old meaning. Science is replete with completely unfamiliar combinations of letters – neologisms or words with Latin or ancient Greek roots. I do not know these languages; at the same time, I want to use words from my native Kazakh, though Russian is also native for me.

I must admit that I also want to draw attention – first of all, of the Kazakhs themselves – to the fact that the people with a small population have a great language, which is not inferior to world languages in terms of its richness and development. After all, the richness of a language is rich intellectual content of the people's life and history. Therefore, I very much hope that these terms will be accepted by the scientific community and take root in the scientific lexicon.

Unlike the concept of *sulde*, which has its physical place in nature, we have come to the concept of *aser* by metaphysical cognition; for us, this is an abstract concept. In Kazakh, aser means influence.

So, we introduce a dialectical pair 'inertia versus aser' in the simple microcosm of the amions. Newton came up with a similar pair – the dynamic force against the forces of gravity – for the complex physical world 300 years ago.

Two Forces of Newton and the Primordial properties of Amions

In the history of science, there are turning points of epochal significance. One of them is connected with the scientific achievements of Newton, who discovered two forces of different nature.

Usually, he has the merit of creating the mechanistic principle and discovering the forces of universal gravitation. However, the discovery of the inertial property of bodies by him and Galileo is no less important for science. Like Newton, modern physicists do not recognize the inertia of material bodies as a force; they call it the **principle of inertia**, or **inertial property**.

Adhering to a different opinion for some time, I believed that inertia is the kinetic force of bodies and particles, the only source of energy in the universe. Then I began to divide complex physical concepts into components, and it turned out that the dynamic force **ma** is the derivative of inertia **mv**. It was my personal turning point.

The dynamic forces of Newton **ma** and the forces of gravity are a dialectical pair in celestial mechanics and the macrocosm. I assume or postulate that inertia is manifested in the microcosm (micromechanics) among the amions. In this case, it must have its own dialectical pair; that is, the property of amions' attraction to each other **mv** – the aser. Like the property of inertia, aser will be considered an a priori property of amions (the primary matter). We are forced to assign the aser property to amions against their inertial properties to create a dialectical pair in micromechanics (quantum mechanics).

Thus, we will record **the following postulate**: as a substrate or the smallest indivisible particle, amions have a normative non-zero mass and two primordial properties corresponding to this mass, or a priori properties – **inertia**, or motor activity, and **aser**, property of attraction, which acts at a distance.

Aser has a quantitative expression **mv** – the aser rate; at the same time, the gravitational forces **ma** are its derivative. For accuracy, we note that the forces of gravity attract physical bodies with acceleration; unlike them, the aser acts without acceleration between the amions. In this sense, the aser property of amions is like an invisible connection, which can both break off and stretch. Aser and inertia of amions are the essence of the primary matter; they are two contradictory properties that constitute a dialectical pair. At the same time, it is important to emphasize for further conclusions that **they are not forces in the classical sense**.

In macromechanics (interaction between physical bodies), Newton's dynamic forces **F=ma** constitute a dialectical pair with the forces of gravitation, and in micromechanics (interaction between particles), they are formed by inertia and aser. Note that the classical definition of the concept of inertia as a fictitious force leads us away from the correct understanding of things.

We will refer to these fictitious forces (a notion from the classical mechanics) as **reactive forces**. Reactive forces arise when the inertial state of both physical bodies and amions changes.

Newton's gravitation laws do not apply in the microworld. This means that, at the level of the microcosm, we abandon the principle of the inverse dependence of the forces of attraction on the distance between the amions.

The forces of attraction between the amions do not depend on the distance between them.

Classical physics tells us that inertia and mass are inseparable concepts; the mass is the measure of inertia and vice versa. Aser and mass are also inseparable. The properties of the aser and inertial property are the measure of mass; the two properties are interconnected through the mass as through a single material entity.

Like Euclidean geometry, an amion with its original physical properties as the substrate is the starting point of the new beginnings of fundamental physics. The amions have their own inertia and aser rate corresponding to the mass Inertia and aser, which are inseparable like the two sides of the same coin. The inertial property is the normative amount of motion (inertia) of the amion and the aser property is the property to change its motion (inertia) normatively. Inertia is the amount of motion **mv**. If we increase the velocity of the amion, we will get the dynamic force **ma**. Aser is the attractive property of amions; in macromechanics, it becomes the force of gravity **ma**.

In our postulate micromechanics, amions in the primary evolutionary medium interact with each other only by inertia and aser; **there are no forces in it yet**! The forces are absent in the Newtonian understanding of their nature and essence. In micromechanics, there is an interdependent motion of amions; forces and acceleration appear only in a mechanical system in cases of transformation of complex dynamical systems. The forces **ma** appear when there are contradictions in the system. Until then, the amions move only by inertia, and curvilinearly –

because of the aser's influence, the value of which does not depend on the distance between the amions.

In the chapter, "The Formation of the Solar System," I will explain how the reverse dependence of the forces of gravity on the distance between physical bodies arises on the basis of the aser's constancy.

To consolidate the new information, let us consider all of the above on specific examples. In Fig. 1, which is already familiar to us from the first section, the dynamics of the planetary system is balanced by the dynamic forces +° of inertia and gravity forces +° of aser.

The plus sign (plus with a mark) means abutment, not addition. It will temporarily serve us in this graphic form in the framework of this book. It is the sign of the interaction of amions' inertia and aser, or the sign of **the vortex beginning of the world**, to which a special section will be dedicated. The vortex beginning, just like the definition of the substrate, is the key point of the formed postulate (see the section "Vortex Formation"). For now, let's assume that the planetary system initially existed as a circular vortex.

The interaction of the dynamic forces and forces of gravity creates a pendulum movement – an eccentricity, as a result of which the planet moves along an elliptical trajectory. If there were no increment and decrease of the dynamic force, then the planet would move by inertia along an ideal circular trajectory. Such a scheme of motion is possible only theoretically; in the real world, it will never be stable; it is like trying to put a vertical needle on a steel surface.

In our world, the nature of the dynamic equilibrium is very tricky; it has the form of a pendulum swinging unnoticeably. The imposition of the dynamic force +° on the planet's inertia with the help of gravitational forces and the addition and decrease of dynamic forces create the pendulum stability of the system. This is a directed and dosed oscillation of the rate of inertia of bodies or an organized group of particles.

In this sense, the structure of the planetary system and cycling are of the same nature, which is called dynamic stability in the pendulum variant. Cyclists constantly turn the handlebar either to the right or to the left to ensure dynamic stability when riding. It is the necessary and unobtrusive eccentricity of the handlebar oscillations that makes it possible to ride a two-wheeled vehicle.

Any vessel on water is always rocking; therefore, to maintain balance in rolling, a special way of walking with great eccentricity was invented – pendulum walking. An observant person will always catch sight of a duck-legged man among the crowd. To keep balance in a swinging medium, the sailor is forced to be swinging himself.

In a dynamic world, only a dynamic balance is possible.

Let's take a different view of this and review the antithesis: will the bicycle without motion (in static) stand on its own in an upright position? Of course not. Our world is arranged in dynamics; the 'bicycle' of our world is always in motion. A dynamic world, like the bicycle, can keep balance only in a swinging model. This is defined (ingrained) at the level of the primordial properties of amions. We can fix this thesis as another postulate.

Aser properties remain constant, and the speed of amions can increase or decrease against the rate of inertia m^1v^1, where v^1 is the standard amion speed. In the known world, the speed of light is the highest possible. This means that the speed of amions can increase up to 300,000 km per second. I will assume that the amions that make up the light 'borrow' extra speed from other amions. The amion can change both the amount of motion and the trajectory, while it always tends to return to its original state. The amion's pendulum swings around a certain zero point, its original inertial state.

Thus, we come to the conclusion that Newton was right; in the planetary system, there is actually a factor of 'divine impulses.' Newton's 'divine impulses' is the adjoinment of the dynamic force to the initially uniform inertial motion of the planet. Why do we use the word adjoinment, or abutment $+°$? It is simply because we cannot add **mv** and **ma**; hypothetically, they can only adjoin.

The average speed of the planet is constant; at the same time, at any given time, it moves a little faster or slightly slower to maintain the dynamic balance of its orbital motion. To keep balance in this swinging world, the planet is forced to possess the corresponding properties and to swing itself.

Medium as a Physics Category. The Concept of Ether

In the IV century B.C., Epicurus and Democritus introduced the two concepts – the vacuum and atoms. Later, physicists began to call it a discrete medium. Discretion (from Latin *discrētus*, meaning 'separate') is a discontinuity that is opposed to continuity. A discrete medium is a discontinuous medium; its antithesis is a continuous medium – the ether.

Further in the book, we will consider different types of medium. The evolutionary medium concept is basic for the general theory of evolution. The last section of this chapter will be devoted to it.

Figure 2 shows a **discrete medium** consisting of air molecules. The forces of attraction between them are very weak; therefore, each molecule moves predominantly by inertia. They are almost independent of each other; in this figure, we see typical chaos. From the standpoint of evolutionary logic, the discrete medium can be considered as an evolutionary system with a high level of active principle: the forces of attraction are too weak to affect the trajectory of inertial motion.

Figure 3 shows a **non-discrete medium** in which elementary particles (amions) bound by the aser property affect each other because of how their inertial motion is displayed in the form of curved lines. The state of the non-discrete medium depends on the density of the amions. When it is high, the trajectory of the amions' motion will be strongly curved, whereas at low density, it will be closer to rectilinear. The change in the trajectory does not affect the aser rate; the aser value remains unchanged at any density. In the non-discrete medium, the properties of inertia and aser interact with each other. I call this medium the **new ether**.

In science, there are words that are no longer widely used: phlogiston, caloric, ether, etc. The word *ether* proved to be more durable; it is widely used in technology and the media to designate some sensitive or broadcast space.

Ether is a previously assumed universal continuous stationary medium, which fills the entire world space. For example, for Descartes, the ether was a continuous medium, a kind of matter smeared in space. At the time, the concept of ether was necessary for explaining electromagnetic and gravitational phenomena and the interaction of bodies at a distance.

The new ether, or the amionic medium, which I also call the amionosphere, is a discrete (intermittent) medium because it is composed of particles, while at the same time, it is also a non-discrete, continuous medium.

The amionosphere is able to pass waves through itself; it is also the bearer of gravitational forces. Perhaps, we have used too many names for the same concept – the ether, or space. These names reflect different facets. Further, practice will show which of them will catch on. I have no doubt that this will happen, although one may ask why introduce a new term if there is an existing one, the meaning of which includes the main property – to communicate at any distance 'in a non-contact way;' that is, transmit all kinds of signals.

Why does the new ether have such a dual nature? The 'old' concept of the ether assumes wholeness (closedness - continuity), homogeneity, and immobility, while it is these properties that the new ether does not have. The updated concept of ether is based on the fact that it consists of material particles – amions – so it can be mobile and have different density in different places.

The presence of two contradictory properties in the pre-atomic matter (amions) makes the medium both sensitive and mobile. The presence of mutual influence (attraction) in amions makes it the world's tremulant web – the new ether. The new ether can transmit all kinds of electromagnetic waves. Simultaneously, the amionic medium transmits – by gravity, from one particle to another – the gravitational forces of physical bodies in a non-contact way. As a result, the properties of amions make this medium a continuous medium; that is, the ether.

Let us touch on the concept of vacuum. The philosophers of Ancient Greece, Epicurus and Democritus, taught that everything around us consists of the smallest particles – atoms – separated by vacuum. In my opinion, there is no ideal vacuum; there is only emptiness between the amions.

Nor is there vacuum in space. Cosmos is an ether consisting of amions, through which information comes from distant galaxies. We make such a conclusion after clarifying the concept of the non-discrete medium, the definition of the ether's new nature. The primary non-discrete medium, or ether, consists of amions – pre-atomic matter.

Now let's get back to the beginning of the section where we considered the concepts of a discrete and non-discrete medium.

Figure 4 shows a billiard table with balls. If we hit one ball with the cue in the direction of a cluster of others, only some of the balls from their total number in the cluster will come into motion. This is what we call a discrete medium. This is not a continuous medium.

Figure 5 shows balls connected together by elastic threads; they are also attached to the sides. After a similar hit by the cue, all balls will begin to move and the sides will deform. We called such a medium a non-discrete medium or ether.

To better reveal the dialectical side of the issue, I propose to consider the third fictional type of medium. Let us imagine that particles in a fictitious medium possess the properties of attraction and do not possess the properties of inertial motion. In this case, all of them will eventually gather in one place. Physicists, without realizing this, present the world as this type of medium. To prove the thesis that inertial properties of material particles interact with the properties of aser, I use a contrario reasoning and apply the antithesis.

Let us consider the fourth, also fictional, type of medium, in which the particles only have the property of inertial motion. Such a medium will gradually expand and lead the world to heat death.

The dialectical side of the issue proceeds from the understanding that by giving the particles the primordial properties, we balance the world. By doing so, we solve more accurately the great mystery of the eternal movement and the vitality (creativity) of complex matter. This is the notion of dynamic symmetry in physics, which is arranged by the properties of the amion. Unlike the aser, the amion has the property of inertia, while the aser must microscopically prevail over inertia. Otherwise, the complication of the world would be impossible; all the complexities in the world would be unstable.

Concurrently, we can conclude that the modern concept of an expanding universe is wrong. Most likely, our universe is just swinging. Then how do we explain the Hubble's 'redshift'? One of the explanations is that we can assume that the galaxies are gradually shrinking and, therefore, seem to be moving away. At the same time, they must be unclenched, so it is necessary to find 'ultraviolet displacement.'

I'd like to continue this section with the thesis that the ether passes into the physical body; at the same time, it is the continuation of this body. Complex structures, starting with the atom, form an amion cloud around themselves – the ether – therefore, all qualitative changes within complex structures are reflected by waves, for example, the ones caused by a stone that has fallen into water. A small stone can cause a fluctuation in the water mass a thousand times greater than it. If we hit a piece of metal with a hammer, our blow will create waves in the metal's protobody. If we act on a piece of metal with electromagnetic radiation, we will also create waves in the metal's protobody, which we call electricity. The protobody of a solid substance and a liquid is a tremulant non-discrete medium, but it will be a complex medium of atoms, where the interatomic medium is composed of amions.

Thus, we arrive at a conclusion that may seem strange: the space surrounding us is both a discrete and non-discrete medium. It is discrete because it consists of amions. The discreteness of space is characterized by its density – entropy – which is uneven; there are many amions in one place, while there is little of them in another. At the same time, space is not discrete and is a continuous medium because amions have the properties of aser and inertia; they affect each other. This makes the medium ether – a tremulant, waving web. In their totality, the properties of a medium may be closer to discreteness or to non-discreteness. One can say they are swinging, just like the entire existing world.

Similarly, disputes between ether theory supporters and their opponents, which flared up in the early twentieth century, resolve in favor of both sides. Our science is also swinging from side to side, and the truth lies somewhere in the middle.

Each body has a gravitational field around itself: I call it ether, or an amion cloud. This field (ether) is a continuation of the body. The meaning of the high word *aura* (of a person) has a physical basis, which includes subtle vibrations of the medium at special frequencies, coming primarily from the brain. The brain itself directly 'peeps' into the world with the help of the eyes and can radiate thoughts and feelings through them, which are often understood without words, directly, without additional encoding by means of words.

In conclusion, I will note that our society is also a medium. The subjects of the social medium are people. Gradually, they become more independent of each other, and this medium deviates ever closer to a discrete type. Philosophers call this process the **atomization of people**. Nevertheless, the society is developing dynamically. People have created a new kind of ether (a non-discrete medium), which is called the information space. But according to the old human tradition, the 'information ether' is used not only to unite people with planetary intelligence, but also to conduct information wars.

Information wars are a consequence of the contradictions between different cultures and ideas. In dialectical evolutionary logic, sooner or later, such a contradiction should lead to the formation of a new, unified world. I'm sure we'll find out very soon whether it will be that way.

Vortex Formation

In this book, I try to formulate the basic provisions of the postulate for the new beginnings of physics, verifying their consistency by creating the hypothesis of the formation of the solar system and the evolutionary theory of climate. Altogether, this is only the first scientific steps toward the creation of a general theory of evolution necessary for further social progress, the development of science and, in particular, solution to the global climate problems.

In the formed postulate, we have already recognized that the smallest indivisible particles called amions have two properties – inertia (internal motor property of amions) and aser (attraction). The aser value is stable and does not depend on the distance between the amions. We have also recognized that the atmosphere and space consist of amions and, therefore, are ether.

The combination of two primordial properties of amions in favorable situations and under certain conditions can create spontaneous vortex of particles in any medium, even in the protobody of solids. A more complex and stable structure is formed only in the form of vortexes of 'restless' amions. The aser, together with the property of inertial motion, cause the vortex motion of microparticles.

In my youth, I watched the herdsmen chase a multitude of horse herds into one place. The required number of additional riders and shepherds together began to advance in one direction, moving in a large circle. The herds of disorderly, almost wild horses, walking throughout the steppe, would then move in circles together with them. The herdsmen gradually closed in the circle to full compaction, gathering horses in the intended spot. If the herdsmen drove the herds separately, they would scatter in different directions.

The same method is also used during the battue, when game is to be gathered in one place. Wild animals and horses, like amions, tend to chaos, but if their unbridled energy is directed in a circle and in one direction, then they are organized into an endless circular motion.

In this case, the herdsmen and hunters know how to manage the wild energy of animals; they act with the knowledge of the laws of nature. In the case of horse herds, the guiding and concentrating power is the consciousness of people: the shepherds were able to direct the kinetic forces of horses into an endless vortex movement.

In natural vortices, the aseric properties of amions act as the concentrating or organizing principle. As a result of the interaction, which I conditionally call **tapping**, the inertial motion of all the amions is organized into a vortex motion around the empty center. We often see vortexes with an empty center in nature. After the amions concentrate with the help of aseric properties, their inertial properties **mv**, or vectors of amion inertia merge into one circular direction. An integral effect is created and the vortex motion is organized.

This is another addition to our postulate.

Evolutionary Medium

The concept of an evolutionary medium is universal for any evolutionary system. It will be common, true, and valid for all its subjects: physical, biological, and social. The evolutionary medium is a concept common to all science; through it, we unite cosmology, physics, biology, and humanities into one whole – the general theory of evolution.

At the end of the first chapter and the postulate part of the book, I will briefly outline the general approach to this concept. We will consider in detail how it is implemented in practice in the next chapter using the example of the formation of the solar system.

We already know that a medium consisting of particles can be of two types: discrete and non-discrete. But only a non-discrete medium often becomes a self-developing evolutionary medium. What is needed for this? The primary evolutionary medium must have additional properties in comparison with the ordinary one. I distinguish four basic properties.

1. *Equality.*

The medium must be homogeneous and consist of only amions.

2. Lack of structuredness.
The density of the medium must be fairly uniform.
3. Autonomy.
The influence of external forces should be weakened.
4. Fullness.
The amount of total mass of amions should be sufficient for development.

To these four necessary conditions, we can add the fifth – the favoring of the external medium.

Almost all amionic media known to us are structured, or, more accurately, any medium is a local site of some more complex material structure. A complex structure is characterized by a dense center (core) and a less dense periphery. The density of amions at different sections is very different, but in general, it varies gradually from the center to the periphery. This is what makes the structure stable.

An antithesis to such a notion of the complex material structure is the modern understanding of the structure of an atom and its electrical equilibrium between a positively charged core and a negatively charged electron.

In the physical sense, the property of the evolutionary medium is expressed by a favorable situation for the generation of vortexes; this situation occurs very rarely in a medium with a developed structure. Why is it so? The primary evolutionary medium as the simplest structure has no laws; there is only chaos. All participants in this chaos have a priori properties. In the evolutionary medium, amions become the only and equal subjects of evolution. Such a medium has the potential for global evolution.

Chapter 2. THE FORMATION OF THE SOLAR SYSTEM

First Stage of the Physical Evolution of the World

In a quiet corner of space, equidistant from neighboring galaxies, in places with particularly weak gravity, there is a gradual accumulation of a large mass of 'stray' particles, or amions.

We should not doubt the existence of such reaches in the galactic space; it is very large and there always will be attractive sites with low entropy for the gathering of space wanderers – they are always attracted to where the influence of external forces is weakened. For example, technical thermodynamics consider the places where sand and other deposits usually accumulate.

Previously, in the postulate part of the book, the only subjects of the evolutionary medium assigned metaphysically were **amions**. Two properties of particles – **aser** (influence, attraction) and **inertia** (kinetics) – are separable only speculatively. To realize the motor activity of amions in pure form, that is, for their translational and rectilinear motion by inertia, a discrete medium of absolute vacuum is required. There is no such place in the Universe since there are no necessary conditions for the complete independence of the particles from each other.

The emergence of the solar system is associated with the emergence of a primary evolutionary medium in space. Amions accumulate between the strongest (energy-wise) galaxies in a place equidistant from them, or in the **center of non-gravity**. They are brought here by cosmic rays.

We know the concept of the center of gravity. Let's introduce its antithesis: the center of non-gravity will be called a 'quiet place,' where the gravitational forces of the galaxies are weakened as much as possible. It was in such a place that a new amionic medium was born, in which the density of the particles (amions) was practically uniform. If we characterize the state of the given medium at zero time using the usual modern language, we can say that there was a creative situation: the opposition was wandering, so there was an 'ideal democracy' of the amions. In galactic space, particles first accumulate in centers of non-gravity, which are then transformed into centers of gravity.

Figure 6 schematically shows a new cosmic formation being formed between the three clusters of stars – a cluster of amions. What is the distinguishing feature of this amion medium? It has a rare feature: no one exerts pressure on it from outside. In this medium, all subjects can fully realize their properties.

Gradually, the new evolutionary medium forms **a large cosmic ball**, which then forms **a large cosmic vortex.** Many cosmologists and evolutionists believe that the shape of the original vortex made it look like a large bagel with an empty center. The astronauts' water balloon was shown to the whole world. By analogy, I suppose that the cosmic vortex twisted exactly in the form of a ball.

The new space formation first forms the organizing center of gravity, and then the self-contraction of the discrete medium under the influence of the forces of attraction occurs. At the first stage of physical evolution, the amions continue to be chaotic, but there is already a general tendency of particles to gravitate toward the center of gravity of the new formation.

Most likely, the specified accumulation of particles in the form of the formless mass is the **dark matter**, which remains a mystery to us until today. It is a conglomeration of pre-atomic matter (amions) in a distant cosmos. It has no form, but already has a gravitational influence in its medium. There should be many such places between galaxies. Later, when the dark matter as a formless formation begins to be structured, to take shape and boundaries, it becomes **a black hole** that readily swallows cosmic rays, but does not radiate them.

Earlier, we adopted the principles of Newtonian mechanics as a theoretical basis for creating the postulate of new beginnings of physics, making adjustments in their action for the microworld, or micromechanics. In particular, adapting (updating) the concept of inertia, we initially accepted the thesis that amions cannot be in a state of absolute rest. In the distant cosmos, the atom, the nucleus of the atom, and as part of the amionosphere, they always tend to a uniform and progressive inertial motion.

In Newton's macromechanics, physical bodies, as well as the smallest particles, can be in a state of inertial rest. According to Newton's second law, $\mathbf{F = m \times a}$. To set resting physical bodies or particles in motion, we need to attach external forces to them.

Answering the question: what is the carrier of these forces, physicists assert that energy is a separate substance (which itself carries); that is, in addition to material particles, it is necessary to search for another kind of energy matter.

According to the inertial rest property attributed by Newton to physical bodies, energy (the property of bodies' motion) is an external substance with respect to material objects. Physics assumes that they can be in a state of immobility; in other words, material objects do not have kinetic energy (an intrinsic property for movement) or an internal motor drive. At the same time, science recognizes that they all have the properties of mutual attraction. Material objects have an internal magnet, but the **motor energy** of the whole world is somewhere apart from matter.

Let us suppose that all material objects have only the property of attraction. Then all matter in our world must accumulate in one place. Let's ask the next question: is there any rest in the visible and invisible world? We find out that all material objects are in inertial motion, so even relative rest is impossible. For example, a stone lying motionless on the surface of the Earth continuously reacts to external influences. It heats up, cools down and, as a result, deforms. In addition, the stone consists of atoms that continuously vibrate and produce radiation. A stationary stone can only have conditional rest; it's not even Einstein's relative rest. The rest of objects can only matter in specific mechanical calculations. The stone is part of the general cosmic motion of the Solar System together with the planet. At the same time, atoms and the tiniest particles inside them are also in continuous motion.

Recall that when constructing a hypothesis, we simultaneously analyze and check the postulate basis of physics that we accepted in the preliminary form. In particular, checking the postulate on the properties of amions, we came to the conclusion that they are in continuous motion, or rather, they possess the property of continuous mechanical motion. Accordingly, they cannot be in a state of inertial rest. On the contrary, their inertial properties are their property of mechanical motion.

In the microworld, Newton's gravity laws do not work, and the forces of attraction do not depend on the distance between the amions. Newton's famous formula on gravitational forces is only valid for physical bodies. The energy source of the entire universe is the kinetics of the particles (amions) **mv**, or their inertial motion. We are talking about a source of energy, and not about the energy itself. For kinetics to become energy (work) in its modern usage, **mv** must first become the force **ma**, and only after this will the specified force perform the work.

Moving forward and in parallel as part of the cosmic sphere, the particles can approach each other, and they can get very close. Amions begin to move continuously and endlessly in a circle in one direction. Figure 8 shows that only with the second option is there a congruent combination of two contradictory properties. Such a combination further leads to a vortex motion of the amions, or to a vortex of the non-discrete medium.

It was Rene Descartes who first proposed the idea of the vortex cluster of particles. Vortexes that arise in a hot desert or in the world ocean, a spontaneous vortex in the atmosphere is not a vortex of air molecules; it is a vortex of the smallest indivisible particles that we call amions. The molecules of air are only carried away by the vortex of particles; they cannot initiate an independent vortex because the molecules of gases are weakly interconnected by forces of mutual attraction and constitute a discrete medium. See Fig. 2.

The binary opposition and a similar combination of forces (properties) in nature have been known for a long time; the dialectical principle of combining two contradictory properties or two principles in one subject is the basis of modern philosophy. In Western philosophy, it is formulated as the principle of unity and struggle of opposites.

What is the difference between the traditional and new beginnings of physics that we are developing in this book? The fundamental difference is that in the new principles, substance (matter) consists of one kind of particle – the amions, but at the same time, these particles have two conflicting properties.

We get a similar idea of structural balance in the macrocosm when observing nature and conducting the corresponding analogies. Let's assume for a minute that amions are mercury atoms. If the mercury atoms did not move and simultaneously attract each other, they would not be able to create new balls almost immediately after a mercury ball is cut. In this example, we see that two forces always work together congruently – one force helps the other.

Mercury as a substance is a non-discrete medium. In the case of a cut ball, the forces of chaotic movement of atoms help the forces of attraction and matter to quickly organize themselves into new balls. It is the combination of two forces in one atom that gives an integrated impulse to the creation of mercury and helps it maintain the convenient (adaptive) shape of the sphere in a given medium. The same properties are possessed by amions, but these properties are many orders of magnitude stronger with respect to their mass.

Experiments in this direction were first conducted by Henry Cavendish; he used a rocker with two suspended objects to which he connected heavy lead balls. These balls twisted the thread only slightly because the gravitational forces were so weak that they were slowed down by the forces of the thread's resistance to twisting. Unlike Newton's gravitational forces, the amion's aser property is original. For example, electromagnetic forces are a manifestation of the original aseric properties of amions in certain situations.

The electromagnetic forces of a small piece of magnet can raise an object, the mass of which is many times greater than the mass of the magnet itself (an object with iron content). The magnet wins over the forces of gravity. In nature, aser properties are manifested only under special conditions because the world is ordered, stabilized, and is in an inertial state.

The self-organization of the primary matter in the evolutionary system is initiated by its internal forces of mutual attraction and the kinetics of the particles. The combination of two forces leads to the formation of a cosmic vortex. Each individual particle has internal forces, or a priori properties. Figuratively speaking, there is a microscopic 'magnet' or motor inside each particle.

These two properties are completely compatible; at the same time, the particles can move in one direction and approach each other very closely. The forces of attraction of particles concentrate in the primary matter, and the forces of kinetics form a continuous circular movement of the particles. As a result, a complex dynamic structure is formed – a self-rotating large cosmic ball (vortex).

The main characteristics of a complex material structure is that it is formed as a dynamic system; the primary subjects of evolution, or amions, realize both of their properties to the fullest extent inside it. In this medium, they continuously move in one circular direction while being continuously attracted to each other. The two properties together create a systemic property of the whole cluster of particles, which in the future will initiate the evolutionary development of nature. Usually, there are no 'on the mooch' amions in stable complex material structures.

Stability of the entire structure is provided by gravitational forces, which maintain the kinetics of all participants in the planetary system. In a complex structure, the aseric properties of individual amions are compiled into the gravitational forces of the entire aggregate of amions. We state the birth of the first physical force, gravity. It is born when the aseric properties of a multitude of amions are woven into a web of complex structure and directed in the form of rays emanating in all directions of space from the center to the periphery of the sphere.

In any complex structure, there is a hierarchy; some participants are subordinate to others. Sometimes in nature and society, there is a situation of anarchy and subsequent short-term democracy. During such periods, there is a restructuring of complex structures and entire evolutionary systems.

In my hypothesis, I view a combination of two properties of amions in a non-discrete medium as an evolution engine. In reality, a situation in which they would move by inertia rectilinearly and progressively is impossible. The aseric property allows only their curvilinear movement. At very low entropy of the medium, or low density of amions, their motion will be close to a direct trajectory.

When the space sphere was large and non-compact, they could move almost rectilinearly, parallel to each other. As long as the amions have a lot of space (environment) and are not cramped, complete harmony will temporarily exist in the new system.

I should note that in the real world, the state of any medium can be closer to a discrete or to an ideal non-discrete medium. This means that there is no perfectly discrete medium and a 'dead' (immovable) non-discrete medium. Any state of the medium is relative.

In the amionosphere (non-discrete medium), the smallest particles are interconnected by the aser properties and simultaneously perform a chaotic inertial motion. Air molecules are not bound by forces of attraction; abstract air is a discrete medium. On the contrary, if air was a non-discrete medium, then it would turn into a liquid under the influence of gravitational forces. At temperatures close to absolute cold, molecules of any gases organize in it.

A medium with low entropy and temperature close to absolute zero is an extremely depleted amionic medium. In a medium with very low density, the weak forces of gravitation of gas molecules come to the fore: gases are transformed (sublimated) into an aggregate state of the liquid.

Molecules of air (gases) in the atmosphere cannot spontaneously initiate a vortex of the medium. At the same time, any liquid will easily do this because its molecules are connected by forces of attraction.

In the old hypotheses of the formation of the Solar System, scientists take only one kind of force – the force of gravity – as the incentive and engine of evolution at zero time. The super concentration of matter in the center of the solar system is explained by the peculiarity of the forces of gravitation and their increase with decreasing distance between the particles.

At the same time, the other side (feature) of the forces of gravity, namely, their ability to increase as the particles (bodies) approach each other, which is to collect all the matter of the universe in one place, is not considered. There is no such phenomenon in the world, which means that matter also has other properties that interfere with the forces of gravity and counterbalance them.

The sun is a massive center; planets and other cosmic bodies revolve around this center. This is nothing but a vortex. In addition, in this rotating space system, there is a primary matter – a collection of the smallest elementary particles, which also rotate around the center. All this together forms a single structure, a large autonomous space organism in the Galaxy.

The last condition needs to be explained separately. Why didn't the vortex start much earlier when our system had a low mass? In this case, there wouldn't have been life, and evolution would have been different. The fact is that the decreased influence of galaxies does not mean that it does not exist. On the contrary, the forces were optimal for accumulating the required mass of amions in a given place, which I called 'a quiet corner of the cosmos.'

External forces held back the birth of a large vortex until the right moment; only after the accumulation of a sufficient mass that the gravitational forces let the new formation travel on its own. The emergence of a new world requires a favorable environment – the galactic system.

In the new history, a similar situation took place in North America. England was far away; its central forces were too weak to control the situation in the region. Against this background, the creative energy of the colonists resulted in the self-organization of a new country; its social structure was built from scratch. There emerged an evolutionary medium (the situation of democracy in the social medium). Theoretically, any colonist had a chance to be president. However, it was only after the accumulation of a sufficient number of colonists in North America that conditions were ripe for the emergence of the United States. This was promoted by the forces of the metropolis: they could not control the situation in the colony and let it go, as is customary to say now, when it became self-sufficient.

At the first stage of the physical evolution of the world, the forces of attraction continuously compact the medium, and the forces of kinetics create a circular form of particle motion. The combination of these two forces led to the creation of the first bodily form of matter – a rotating sphere.

Formless and chaotic accumulation of particles in the galactic space (disorder) as an evolutionary system (medium) randomly self-generate first bodily formation of a circular motion of the particles similar to a steppe vortex. The newly formed sphere has the first materialized boundary – **crust** – similar to a film on water. The highly entropic material (rich) sphere separated from the low-entropy (poor) section of the galactic medium by means of the crust.

It should be added that the first vortex in space occurred in the primary matter medium, when there were no atoms yet. To consider the cosmic dust and gases as a 'raw material,' a substance, and a vortex substrate is an omission by evolutionary scientists, just like disregard for the inertial properties of particles.

Why do modern scientists not accept elementary particles as part of the first vortexes? I think it's because then it will be necessary to explain how the atom was born in the Solar System, and this task is even more difficult than revealing the secrets of the origin of evolution. It is not easy to offer a coherent theory about who, how, and why someone put electrons and protons inside the atom. For such a construction, we will first need an 'assembly factory,' and secondly, some reasonable power. To find out how the atoms were created in the process of general evolution, it is necessary to consider various versions of their construction. We are creating (constructing) an atom from one primary element with two properties, and modern physics is made up of two or more primary elements.

Cosmic dust and gases in the form of semi-finished products (atoms) cannot spontaneously evolve to the state of solar plasma. To explain this using existing hypotheses, scientists sought external reasons or some kind of a cosmic accident. Some external forces were supposed to 'ignite' the dense substance of the protosun; scientists needed a 'fuse' to begin the processes of a thermonuclear reaction. The introduction of external forces for the initiation of thermonuclear processes contradicts the third basic principle of dialectics, which suggests that matter is prone to self-development, and that the processes in the material world are spontaneous.

One can (and needs) to see a vortex beginning in all structures, phenomena, and systemic constructions in the universe, such as the galaxy, nature, and society. The basic concept of cosmology as a science must take into account the vortex nature of the whole universe.

In conclusion of this section, I want to return once again to its starting metaphor of 'a quiet corner of space' and consider the antithesis of 'a non-quiet corner of space.' The surface of our planet is a non-quiet corner of space because there are powerful unidirectional gravity forces of the Earth. In conditions of 'a quiet space' and 'non-quiet space,' the processes of spontaneous self-formation, or vortexes, in the amionosphere must occur with varying degrees of intensity. The first beginning of our world is characterized by the existence in a large space of this 'quiet corner,' since there were practically no external forces.

The evolutionary processes of self-formation of the cosmic sphere in a discrete medium with very low entropy occurred under conditions of ideal weightlessness, and, therefore, could have occurred autonomously. The formation of a 'quiet corner' in the galactic space should be regarded as the emergence of a favorable situation in the evolutionary medium that gives rise to all subsequent processes.

Second Stage of the Physical Evolution of the World

In a quiet corner of space, equidistant from neighboring galaxies, in places with particularly weak gravity, there is a gradual accumulation of a large mass of 'stray' particles, or amions.

In the first chapter, we called the aforementioned accumulation of particles amionosphere, or a new ether. In a new space formation, the organizing center is first formed. At the first stage of evolution, the amions continue to be chaotic, but there is already a general tendency of particles to gravitate toward the center. Figure 6 schematically shows the beginning of the formation between the three galaxies of our solar system – the then new space formation.

Gradually, the aggregate of primary matter – a huge array of amions – continues to self-compact in a quiet corner of space under the influence of its original aseric properties. A large cosmic sphere is formed, which stands out in sharp contrast: it receives cosmic rays, but it does not yet shine. See Fig. 9 and 10. He builds and strengthens a voluminous aseric network of amions, which can be compared to a giant web. It looks like a huge spatial 'farm' inside the cosmic sphere. There are amions in the nodes of this 'building structure,' and the connections of the farm are the forces of attraction between them. A film appears on the space sphere; its self-formation occurs according to the same laws as the formation of a film on water (on a water ball).

In Chapter 1's section on medium, we have already considered an example with a billiard table. If we imagine the space (medium) filled with amions (elastic balls) instead of it, then upon exposure to one amion, everything comes into motion because they are interconnected by aseric properties. These connections can be torn, like, for example, a water column in a siphon break situation. Water, like most liquids, can be broken very easily.

The amionosphere possesses all the properties of a liquid; at the same time, it has a fundamental difference: its density at different sites differs greatly, so I call the amionoscope by its name, and a liquid. It is not a discrete medium in our understanding. Sometimes I can apply this name to it, but only to emphasize its fullness with particles and the lack of structural aggregate content with a dedicated center.

The amionosphere is a volumetric gravitational network, which is the basis of its etheric properties.

The expression 'gravitational network' does not accurately reflect the essence, since the primordial properties of the amions act inside the sphere, and the forces of gravitation appear outside the film, or the limits of the sphere. Therefore, we will call the gravitational network the **aseric network**.

Here, it will be appropriate to criticize the thesis on the four fundamental force interactions adopted in quantum physics. It consists in the fact that nuclear force interactions begin to appear in the center of the sphere, where the density of particles increases, weak forces appear on the periphery of the sphere, and gravitational forces act beyond its limits. Electric forces will appear later – this is possible only in a complex structure of substances.

It turns out that physicists call the four kinds of fundamental force interactions the only one kind of force, arising on the basis of the aseric and inertial properties of amions.

In the modern ('ready-made') world, the amionosphere is a continuation of cosmic bodies. **Inner space** is the interplanetary medium – the prototype of the Solar System, and the **Outer space** is the galactic medium outside the solar system. Both are amionospheres.

It is necessary to note an important circumstance. At the beginning of time, the cluster of amions in a quiet corner of the galactic space we are considering was not an integral part of other cosmic bodies; only hermit particles accumulated and gathered in a new company there. Together, they formed the dark matter, which we discussed above. This statement can be paraphrased artistically: the opposition gathered; a subtle alternative to the well-established and strong galactic world was born.

The main feature of the new formation was freedom from the influence of strong space systems (objects). Only the inner original forces of particles acted; practically, there was no third force; there was no influence of strong centers of complex structured 'neighbors.' In such a non-discrete medium, it becomes possible to accelerate the birth of new structures and even new worlds.

Amions are in continuous inertial motion. The magnitude of the amions' 'run' depends on the distance between them, which, in turn, depends on the density of the amionic matter. In galactic space at the beginning of time, in its quiet corner, the density of particles of the medium increased in two ways.

1. In a quiet and neutral corner of space, particles from neighboring galaxies accumulated.

2. In the center of non-gravity of the galactic space (in the space with practically no influence of star clusters), the self-compacting of the cosmic sphere under the influence of the original forces of attraction of amions occurred.

We talked about the fact that the amionosphere is a continuation of physical bodies. This means that amions are always busy with 'normal work.' In space, they 'serve' the gravitational network of planets or stars. Outside of complex structures, this is a gravitational network, and within any structure, they are forces of attraction. Between the amions, this network will be aseric.

In the evolutionary medium, there is no force network; there are no forces at all, only the aseric network of amions. Forces arise only in a complex structure when the properties of the amions are organized in the form of vectors (impulses of forces) and turned in the same direction.

The amions in Earth's atmosphere are also busy with their participation in the gravitational network; it is the original impulses of the amion forces that create it. When they are 'busy with work' in the gravitational network of any system, their primordial properties manifest themselves only with a sharp compaction of the amionosphere. At the same time, amions unoccupied in the gravitational network – the 'unemployed' amions – sometimes appear in it.

Here we can, again, lean toward the humanitarian field of knowledge and draw an analogy with the people's world. If a lot of idle or unemployed people appear in society, the probability of various kinds of vortexes' formation grows: people inevitably will start new enterprises, which can be either good or not. When the medium is compacted by new particles, there is a disturbance of regular peace (balance in the medium); the original forces of excessive unbound, and then regular amions cease to obey the forces of gravity of the Earth. At a time like this, strong atmospheric phenomena can occur, such as typhoons, tornadoes, and storm winds.

Let's get back to the first beginning of the world. There are quiet places between the galaxies – the centers of non-gravity in which there is a conglomeration of particles. In these places, the gravitational influence of star clusters is very weak; autonomous evolutionary processes, which will be initiated by the primordial properties of the particles themselves, may begin there. Let's recall the behavior of water inside a spacecraft: in a state of weightlessness, it easily self-forms into a ball, as the third force – the force of gravity – is weakened in space. When being descended into a pipe in terrestrial conditions, it inevitably forms a spiral rotation, since the swirling of water occurs when it is in a temporal state of weightlessness. An arbitrary swirling of the liquid and a downward motion form a spiral motion. In other words, two simultaneous movements of water molecules combine in the spiral motion of the jet.

As our cosmic sphere compacts, the initial chaotic motion of the amions within it transforms into a coordinated movement; ***the great cosmic vortex*** occurs – the first launch of the movement of orderly matter.

The process of the primary matter integration or the vortex of the cosmic sphere is much faster than we can imagine. The general chaos of the particle motion is preserved, but there are already tendencies of amions' aspiration toward the center. They begin to simultaneously rotate around the axis of the sphere symmetry – a vortex movement, which we have seen in nature, is formed.

This vortex should have an 'empty center,' which is similar to the eye of any vortex in nature. See Fig. 10. The idea of an empty center is introduced into physics for the first time. It can explain the presence of the Earth's magnetic field and the various properties of atoms and substances. An empty center means that the particle (atom) does not have a permanent place or a conventional spindle rigidly fixed in space. Such structures are constantly vibrating at certain frequencies, which becomes the calling card of any atom and more complex combined structures. It is possible that this is how leukocytes, by nominal frequencies, distinguish the cells of the native organism from foreign microorganisms.

An empty center is formed due to the fact that the amions cannot rotate around their axis. For the forward inertial movement as part of the vortex structure, they need a circular trajectory. This is the only way they can move by inertia.

The directed forces acting from the center gradually transform (subordinate) the chaotic movement of the amions, and then organize the vortex motion around the axis of symmetry or the empty center.

A vortex motion is formed since there is no other alternative; no other variant of motion is possible when two basic properties of matter are combined. There are no other options for constructing a complex structure of amions in the combination of their basic properties that we have accepted. They can get closer, only moving in one direction. See Fig. 8.

The cosmic sphere is compacted by the forces of attraction from the center. The non-discrete medium automatically forms a vortex motion around the empty space – the eye of the vortex. Since the amions are very small, they can later concentrate in the vortex plasma – the center of the atom.

The self-concentration of a cluster of particles (amions) in the form of a sphere and its arbitrary vortex is the first stage of the complication of matter, or the first stage of evolution. See Fig. 9 and 10. The complication of matter must lead to the emergence of new phenomena and processes, as well as the links between them – the new laws, which we observe. All complications in nature are created by nature itself in the process of its evolutionary movement; it is a phenomenon of self-development, a phenomenon of life.

The philosophical definition of life in the form of vortexes has enormous gnosiological significance for understanding the entire history of evolution. As a result of compaction and interaction of amions, the initial processes of self-complication of matter have arisen, and the origin of the Solar System has occurred. In conditions of the necessary level of 'tightness,' or high density of a non-discrete medium, the amions formed an infinite circular movement (vortex) in the form of a rotating sphere.

Once again, getting driven into the humanitarian field of knowledge, let's compare the physical processes described above with the birth of Christianity. In those days, the bearers of new ideas – hermits – hid in such secluded places as deserts, marshes, mountains, and other hard-to-reach areas of the terrain. In Russia, such a place was the Dnieper rapids, where the most obstinate peasants from all over the country were sent. A secluded place is a place in which there is no influence on the part of other (central) forces. When there were too many runaway peasants, they appointed *atamans* (local centers), and then the vortex of the Cossackdom appeared. The free power of the Cossacks played a significant role in Russia's expansionist strategy.

In any kind of movement, one can discern the processes of evolution and de-evolution. Almost all processes – natural (physical and biological) and social – can be imagined as a vortex of the medium and evolution. First, the center of gravity of the dynamic system is created and then the directed forces of attraction arise, which can be called the organizing forces. In the general theory of evolution being developed, I propose to call these forces **an organizational principle**, and the forces of inertial motion – **an active principle**.

Thus, another dialectic pair is formed. In a complex structure, the force of attraction of the center is the organizational principle, and the kinetics of the amions is the active principle. Their dynamic balance creates the prerequisites for the stability of the entire structure. The dialectic of creation lies at the lowest level of the self-organization of original matter into more complex structures. *Integration of the original matter is initiated by the interaction of its two properties, which are the basic cause of all stages of evolution.*

From the point of view of dialectical evolutionary logic, which is based on the ideas of Newton's mechanistic principle and Laplace's determinism, this law also applies in society. Through a very complex series of sublimations, the simple properties of amions are transformed into social ideas. In an individual, the energy of life (active principle) is inherent. In the social environment, there are invisible ties between individuals; they can be family ties, economic, or ideological. The initial social units (such as genus, mahala, hutun or medieval small cities) are a non-discrete medium, an emotional ether. There is an identity between individuals; it is the basis of the patriarchal way of life. As soon as individuals became economically independent from one another, the patriarchal way of society began to collapse.

The dialectic of creation lies at the lowest level, at the level of the primal matter; integration of primal matter is initiated by the interaction of two properties of primal matter; the two properties of matter are the basic cause at all stages of evolution.

Let's sum up the subtotals. On the second stage of physical evolution considered by us, the cosmic sphere begins to separate from outer (galactic) space as the self-compacting begins, since the gap in their entropy (density) grows. This process occurs through the formation of the crust. We will discuss the mechanism of the crust alignment in the Solar System in detail in the section dedicated to the fifth stage.

Third Stage of the Physical Evolution of the World

Now for further construction of the hypothesis, it is important for us to consider the nature of the centrifugal force's origin. Let us start with a simple example. Suppose a car is running at a high speed by inertia. If we sharply turn the steering wheel to the left without reducing the speed, the car will begin to skid to the right. The steeper the turn of the steering wheel, the stronger the skid. There are laws of nature, which we cannot change. The change in the car's inertia gives rise to reactive centrifugal force.

Forces in the general system appear as a result of a combination of the aseric properties of amions and their tendency to rectilinear inertial motion. The motion vector of the amions is inertially directed along the tangent line to the conditional equator, and the aseric properties begin to rotate them along a circular trajectory – **the centrifugal force** arises in the general system. As the sphere becomes more self-compacting, the trajectory's curvature will increase or the curvature radius of the sphere will gradually decrease. According to Newton's third law, the effect of gravitational forces from the dense center of the sphere on the inertia vectors of individual amions causes their equivalent counteraction in the form of the centrifugal force. See Fig. 11.

In the microcosm, the medium of amions, the forces of attraction between particles differ from the force of attraction in the macrocosm. Aseric attraction in the microcosm does not lead to acceleration of amions; it does not depend on the distance between them. Let's also note another point: the centrifugal force is created not directly by the aseric properties, but it is a function of the entire dynamic system. The inertial properties (the amount of motion) of the amions are transferred by the aseric network from the internal rows to the outer rows. As a result, the speed of the amions at the periphery increases due to a decrease in the amount of motion (speed) of the amions in the internal rows.

Further, there is a redistribution of forces in a complex structure: the peripheral amions have an excess of kinetic force, and in the central part, the speed of amions decreases. Therefore, in the center, there is a relative excess of forces of attraction, and they are already becoming forces in the classical sense. However, this cannot continue indefinitely: considering the dimensions of the Solar System, the amions on the periphery must have speed much higher than the speed of light. To prevent this from happening, there is a compensatory mechanism for the entire system.

The aseric network is magical. In this system, congestion is impossible: if one amion is delayed, then the leading amion hooks onto the other. There is also magic in the emergence of the centrifugal force in the system due to a combination of inertia and attraction forces.

As the medium compacts, the angular speed of the sphere is reduced as self-compaction increases because the path (circular trajectory) of the amions is shortened. As a result of this acceleration, centrifugal force is formed in the sphere structure. The spherical form of the cosmic formation gradually builds (extends) into the shape of an ellipsoid.

If we regard the forces of attraction in the dynamic system (in the amion medium) as a dynamic (active) force, like a kinetic force, we find ourselves in a theoretical impasse. Two contradictory and active forces will create a zero balance of energy; a stagnation situation will inevitably arise in the system. In reality, this does not happen. The cumulative kinetics (mechanical energy) in any dynamic system consists only of the sum of the inertia properties of the individual amions **mv**.

We come to an interesting phenomenon. Inertial forces and forces of attraction cannot quench (or more precisely, destroy) each other. The effect of some forces on others gives rise to third forces. In other words, the effect of some amions on others leads to a redistribution of forces in the system.

The effect of the amions from the outer row on the inertial state by forces of attraction from the center brings reaction. Some new forces emerge, but the cumulative amount of motion **mv** in the system remains unchanged (constant). New forces appear as a result of the redistribution of the kinetics of particles: some amions lose speed, while others increase it at the expense of the former. As a result, all amions that are part of a complex structure are subject to the system requirements.

As the primal matter in the autonomous system becomes more complicated in terms of structure, new phenomena arise and the physical laws of nature are born. Before that, we talked only about properties, and now we will talk about laws and forces. They are the result of sublimation of the primordial properties of amions. The centrifugal force pushing out the amions of the outer row in the radial direction arises due to the decrease in speed of the inner row amions. Further, the processes of redistribution of the speed and kinetic energy of the particles are transmitted along the chain to the very middle of cosmic formation.

As a result, the amions in the center – i.e. in the place with the highest density – begin to move and rotate slower than the nominal speed. The cosmic sphere becomes a single membrane system of interconnected forces. See Fig. 12 and 13. This phenomenon requires its own name, which can subsequently become generally accepted, and, perhaps, readers will be interested in reflecting on that matter, offering their version. I would call it the **system springing**.

Figuratively speaking, in the ecliptic plane, this membrane system becomes like a harp: its stretched strings create planar stability. The role of stretched strings is played by the aseric properties of amions; their interaction creates bonds (springing, tension) in the radial direction and along the ring.

In addition, system springing can be compared to a tightly stretched bow; the amions move along the trajectory curvature, which they tend to arch, against their inertial will. The forces of their mutual attraction push the string up and create balance in the system.

The system springing can also be called an effect of systemic development, an increase in the kinetic potential of the periphery, or an increase in the center's attractive potential. All the given names are fair, but they emphasize only certain facets of the system phenomenon: two shifts make the system stable. In addition to this, the occurrence of the centrifugal force makes the system flat. The amionic matter of the cosmic sphere begins to turn into a flying saucer.

Let's try to clear up the redistribution mechanism of inertial properties (momentum) of amions in the cosmic sphere structure. Their contradictory properties led to the self-formation of this sphere and its vortex. The plurality of amions begin to weave laces around a dense ring center; a vortex is formed as a form of movement of the primal matter first in the form of a sphere, and then an ellipsoid.

Why did I come up with the idea of a spring, considering the emergence of the centrifugal force as a function of a centralized system, and the force of attraction the passive force, but a connection? In the conventional sense, the emergence of the centrifugal force in the planetary system is the result of a change in the inertial state of bodies due to the influence of the center's force of attraction. This is true for physical bodies: they are always in a centralized inertial system.

We observe the following phenomenon in the cosmic sphere. Due to their aseric properties, amions are compacted freely on the Y-axis (see Figures 9 and 14), or in a direction parallel to the axis of rotation of the cosmic sphere. As they compact, their inertial state also changes. However, we do not observe the emergence of reactive (centrifugal) forces in this direction. I believe that the forces of attraction between the particles acting in the system along the Y-axis do not perform any work. We can say that they perform a 'fictitious' work. This phenomenon is of a fundamental nature and requires additional research.

To illustrate the above phenomenon, let's imagine a well-known children's toy – a spinning top. It resembles a fantastic flying saucer, which rotates with the help of inertia forces, or more precisely, the moment of inertia. Imagine that our toy first was a cosmic sphere consisting of amions, which gradually (by rotating for a long time) transformed into a plate with the help of aseric properties. If we put a small object on the plate, we will see that the centrifugal force throws it out in the radial direction. These is the real centrifugal force that arises in the ecliptic plane.

We see that as long as the forces of inertia are not exhausted, the spinning top can stand at one point vertically, maintaining equilibrium and dynamic stability of the system in the ecliptic plane. But our 'natural' spinning top – the cosmic sphere – will never run out of the inertial properties. Therefore, the Solar System, together with planets and satellites, is eternally in the ecliptic plane. In the direction to which the centrifugal force does not apply (along the Y-axis), the aser compacts the sphere and turns it first into an ellipsoid, and then into a plate. The sphere becomes a spinning top with a convex center.

One thing is clear so far: the aseric properties become a force and can perform work only under the condition of resistance of the inertial system (IS). Aser vectors acting along the Y-axis are directed vertically to the centrifugal force and the force of kinetics; therefore, they are not subjected to the IS resistance and compact the sphere in the form of a spinning top. The inertial system of self-moving systems is always created in the ecliptic plane. Tools for orientation in space such as gyroscopes are created according to this principle, and, perhaps, the most complex organic molecules are born.

The new concept of IS fundamentally different from the inertial frame of reference adopted in modern physics, but does not reject it. The IS has a complex hierarchical structure, and most importantly, each IS has its own power center. The new IS in the form of a large Solar System also includes local IS.

Now let's try to understand why the centrifugal force appeared along the X and Z axes in the ecliptic plane in the future Solar System. On the X-axis from the periphery to the center, we will arrange a sequence of amions and imagine that they are all connected by semi-rigid bonds (forces of attraction). On the plane, we will visualize the orbits of the amions in coupling as a set of concentric circles. Amions of the outer row move in orbits with less curvature, and those of the inner row in orbits with greater curvature.

With parallel motion, the amions of the inner row 'pull' the outer ones by means of the aseric bonds. Moving along a more curved trajectory, they try to change the trajectories of the amions of the outer row in which the centrifugal force emerges in response to this influence, taking away a part of the inertia **mv** from the amions of the inner row. See Fig. 13. In other words, the excess forces from the inner row are transmitted through the connection of the aseric network, and radial (centrifugal) forces appear in the radial direction in the outer row of amions. See Fig. 11.

Thus, we resolved an important matter: it turns out that the centrifugal force does not create attraction forces. They are only intermediaries in the redistribution of internal inertial forces of amions. This is the fundamental basis of the law of conservation of energy.

Fourth Stage of the Physical Evolution of the World

The rotation of the cosmic sphere, which gradually turns into an ellipsoid, is the coordinated and directed motion of the entire set of amions by inertia, a way of their existence in the form of an organization, and, simultaneously, the manifestation of a priori properties of each individual amion.

They coordinate the directed movement, forming an autonomous vortex structure, in the center of which their overwhelming majority gradually accumulates. Accordingly, most of the forces of attraction of the new formation operate from this center; the greatest amount of kinetic energy of the whole aggregate of amions is concentrated here. The center becomes the leading force, and the concentric circles on the periphery gradually become guided.

As it self-compacts, the ellipsoid accelerates the rotation (angular speed). This leads to the fact that the problem of congruence in the new formation is aggravated. To maintain congruence, the amions on the ellipsoid periphery must move faster, but they lack the normative inertial energy, or their own speed, for this.

Let us consider this situation with the example of athletes' race in concentric circles in a stadium. Runners in the outer rows (at the periphery) will always lag behind the rest at angular speed. See Fig. 15. Like athletes, amions on the periphery begin to lag behind in the general rate of the ellipsoid rotation.

Next, we will analyze how the issue of coordinated movement (congruence) of the cosmic formation is solved. At the same time, we will bear in mind that, firstly, the cosmic ellipsoid is connected by a three-dimensional aseric network (a bulk web) consisting of all the amions; secondly, the amionic medium (amionosphere) is similar to a liquid; third, the amionosphere of the solar system can be represented as a semi-rigid ellipsoid disk. See Fig. 14.

We see that the kinetic energy of rotation of the 'strong' center starts to push the delayed amions at the periphery; it is carried out through the nuclear network. See Fig. 19. The inertial state of the amions, their inertial rest in the general system, is violated, as a result of which **the Coriolis force** arises. I call the counteracting forces that generate the Coriolis acceleration the **pushing forces**.

Let's consider this effect on the example of a grindstone. At first glance, it may seem that the grains of sandstone from the grindstone will fly off by inertia in the direction of the stone rotation, but this is not what occurs in practice. See Fig. 18. The trajectory of sparks from the grindstone is created by the Coriolis force. All objects on the earth's surface are in their IS - a state of inertial rest - including the grindstone. The grindstone and all its grains of sand 'want' to be at rest. The electric engine of the grinding tool begins to turn the stone against its 'inertial will,' and, therefore, as soon as the grains of sand break away from the stone, they fly off against the direction of the forced movement. The French physicist, Coriolis, discovered the forces that initiate such a movement of the grains.

Additional kinetic energy of rotation is transferred to amions at the periphery from the center along the aseric network (along the spring system). On the other hand, the structure of the 'protoatom' is also a viscous medium (liquid), so there is a partial slip of concentric circles in relation to the center and to each other. In any case, each outer concentric circle will be slightly delayed relative to the adjacent inner circle. This effect can be compared with the sliding of the clutch in the gearbox: the clutch gently switches on (transmits) the motor's forces to the chassis of the car.

Figure 19 shows the movement of the amions (runners from the above example) in the coupling. At break of the coupling, the amions (runners) are overturned. This is the basis for the future emergence of internal orbital laws in the Solar System. The outer row amion detached from the coupling falls under the influence of another amion from the inner row. Then the slippage processes are repeated. Impulses of the Coriolis force initiate local vortexes of amions at the periphery in the ecliptic plane of the Solar System. See Fig. 20.

It is important to understand that the central pushing forces, or more precisely, the excesses of inertial forces from the center, transmitted over the network, violate the inertial state of each amion individually and together. Their motion with a constant (primordial or normative) speed is the state of inertial rest. The inertial rest of amions is the rest in dynamics when moving at normal speed.

After they are pushed (accelerated), and their inertial state (rest) is violated, the force of counteraction – the Coriolis force – arises on the periphery according to Newton's third law. For example, if a car moving by inertia is accelerated (pushed from behind), then objects and passengers inside will be thrown back. This is the manifestation of Newton's third law.

If we move a cart with a box filled with water in one direction, then a wooden bar floating on the water will move in the opposite direction. The bar makes a complex movement: it moves forward with the water, while it also moves backwards; that is, it simultaneously makes two differently directed movements. It also happens when a person inside a train moves back, but at a slower speed, and simultaneously moves forward along with the train.

If a viscous liquid is poured into the box, the speed of the bar in the water will decrease. Let's start moving the cart along a circular path. The wooden bar makes an even more complex movement: it moves forward with water, then moves backward relative to the water, and also swerves to the side, moving along a circular path with respect to the water. Approximately the same effect is achieved in concentric circles on the periphery of the entire cosmic ellipsoid, where the amions make a triple motion. Each amion is a bar floating in a liquid, and all together they constitute a liquid. The role of the moving cart is played by the aseric network. Counteraction to the pushing by the system inertial forces leads to a local vortex of the amions at the ellipsoid periphery.

Since the entire moving system tends to inertial equilibrium, the vortexes at the periphery occur in the direction opposite to the rotation of the entire cosmic ellipsoid. This is the **Coriolis Effect**. Vortexes in the peripheral disk part of the ellipsoid gradually lead to the self-formation of planets. For the same reason, they will rotate in the opposite direction in the future. Nowadays, the planets, with the exception of Venus, continue to revolve around their axis under the influence of the Coriolis force. If new counteraction forces appear in the system, local vortexes appear in the concentric circles of the cosmic ellipsoid.

Venus shifted its orbit and hit the neutral zone of the ecliptic, into the zone of strong slip of the disk without Coriolis force. Venus is in a zone of strong slippage in the amionosphere between the spherical and disk parts of the Solar System. Therefore, Venus rotates around its axis, which is tilted 177.36 ° from the perpendicular to the orbital plane from east to west; that is, in the direction opposite to the direction of rotation of most planets.

In its physical basis, the Coriolis force is counteracting the forces that push the system, which is a consequence of Newton's third law. Then why were they given a new name instead of being called a centrifugal force? This is a fair question. In fact, these are the two names of the same force.

Let us consider the nature of their appearance again. Part of the kinetic energy from the center is transmitted through a volumetric network of forces of attraction (volumetric force web) into concentric circles on the periphery of each individual amion. In an ellipsoid, they are all interconnected by this voluminous network. If the centrifugal force arises as a counteraction to the force of attraction of the center, then the Coriolis force arises as counteraction to the forces that push the system.

As a result of their action, the formation of planets occurs.

Amions on the periphery move primarily due to the action of internal (own) forces, while the outer row lags behind the inner row. Because of this, each inner row of particles pushes (or indirectly pulls) the outer row of particles. This idea requires an additional in-depth study and a better understanding of micromechanics as a scientific discipline, which can choose the nature of the appearance of centrifugal and the Coriolis forces and vortexes as its subject.

But let's get back to the original question. Powerful forces of kinetic motion from the massive center through the bulk aseric network begin to push the periphery that lags behind when rotating. The difference between the centrifugal force and the Coriolis force is that the former acts in the radial direction and the latter acts along the tangent. Together, they create a rotational momentum and participate in the rotation of amions on the periphery, or in their vortex. Of the two vectors of reactive forces, a third vector of the new force must be formed, which will be located within the angle of the two original vectors.

Thus, both the centrifugal force and the Coriolis force basically have the same origin – they are the result of a forced change in the inertial state of amions.

In a self-developing dynamic system, contradictions inevitably arise. These contradictions are the engine of evolution.

In the microworld mechanics, they should be combined into one phenomenon. Centrifugal force is the result of the difference in the curvature of the amions' orbits in parallel rows, and the Coriolis force results from the angular lag of the amions in different rows due to the difference in the length of their orbits, as shown in the example with runners in the stadium.

In this chapter, we traditionally consider the above secondary forces individually, because this is how it formed earlier and how it is generally accepted in physics and engineering. The reason is simple: in the macroworld, in our daily practice, the above forces usually manifest themselves separately. In one case, the vehicle moving as if by inertia is pushed behind – and the Coriolis force follows. In another case, a car moving in a straight line can swerve – this is when the centrifugal force appears.

It makes no sense separating these two processes in micromechanics; in the microworld, they occur simultaneously. For this reason, I have an unjustified repetition in the depiction of schemes in the figures. So for completeness, we add Figure 24 and introduce a common vector instead of the two force vectors. When necessary, it can be broken down into two components along the X and Y axes.

The fans of football will easily understand how exactly amions make a complex movement 'three in one.' A football player can strike a ball so skillfully that it will simultaneously rotate and fly along the trajectory curvature. The goalkeeper does not know where the ball will hit, and even if he makes a guess, the ball rotating around its axis is very difficult to catch and return (it's not clear where the ball will fly off).

As a cosmic entity, the Solar System is both a self-organized developing dynamic system and an evolutionary medium. It consists of a myriad of conventional subjects of evolution – amions – each of which has its own regulatory speed. Striving to adapt to the 'habitat,' all particles move by inertia and are simultaneously attracted to each other in a voluminous aseric web. In a large space system, amions tend to create a complex material structure in the form of a vortex, **the Big Atom**.

However, with the gradual compaction of space formation in the evolutionary system of amions, contradictions arise in the structure created by them; congruence is violated. Because of these contradictions, internal vortexes are born inside a large vortex. Vortexes of the third level are born inside the vortexes of the second level, and so on. If we look at these processes from the point of view of evolutionary development, then we can say that a large system (a large vortex) begins to generate new systems, new material structures with a vortex nature, within itself, in which new forces appear and new physical laws are manifested.

The following two points should be noted. In the primary galactic evolutionary medium, subjects were exclusively amions, and in a complex medium, there are new compound subjects of evolution – planets. The second point is that a new phenomenon, a way of evolution – the phenomenal process of **implosion** – emerges. See Fig. 25.

Implosion is an inwards explosion. By implosion processes in the center of the Solar System, I mean successive internal vortexes that can continue up to the level of nanovortexes – protoplasm of future atoms, which subsequently becomes a source of solar radiation. If on the periphery of the Solar System, we see only three levels of vortexes, then in its central plasma part, we can see dozens, or even hundreds of levels.

The plasma miniaturization of solar vortexes inward, or implosion, requires a lot of kinetic energy; the Solar plasma has plenty of it.

We mentioned above that the norm of amions inertia decreases inside the solar disk. Because of this, their relatively intensified aseric impulses ultimately compress the central part of the Sun (the Protosun), the center of which becomes an immeasurably compressed spring. We already know that symmetrical reverse vortexes are formed there (see Fig. 22a), which also participate in implosion processes. They lead to the appearance of the **direct** and **inverse Coriolis force**.

The previously known Coriolis force, or the direct Coriolis force (direct reactive forces), have a dialectic pair – the inverse Coriolis force. The direct force is directed to the creation of planets, and the inverse force is realized in the form of solar radiation.

The symmetrical reactive force creates the highest plasma density and inverse implosion. Nanovortexes are formed in limit states, with the development of implosion inside the 'disk' of the Sun. If on the periphery of the Solar System, the implosion processes expand sideways, then inside the disk, they develop in cramped conditions. The deeper the implosion, the smaller the dimensions of the vortexes and the radius of the vortexes' curvature reduces to the limit. But it seems that the amion can no longer twist the trajectory, so in an instant, the 'spring' of amions fastened by asers straightens and goes off from the core plasma. As a result, the nanovortexes produce the phenomenon of solar radiation. The emission of amionic matter (particles) in the form of solar radiation is continuously compensated by the super-powerful forces of gravity of the Protosun center or the Sun itself, or the forces of attraction make up for the emission of particles by their assimilation. See Fig. 26. This is how atoms 'breathe.' One end of the whip (see the chapter on Evolutionary Theory of Climate, section Implosion: The Stick Effect and Quantum Mechanics) develops to create planets, and its second end begins to work as the energy source of the entire Solar System.

Nothing in the world vanishes and appears out of nowhere. It all boils down to sublimation of forces and implosion.

It is the implosion that leads to the formation of micro- and nanovortexes, which are the source of electromagnetic waves. As a result of implosion processes, the solar plasma begins to emit electromagnetic waves in various ranges: all planets and their satellites located in the ecliptic plane begin to be exposed to solar radiation.

The explanation of the nature of vortexes in the non-discrete structure of an ellipsoid by the appearance of Coriolis and the centrifugal force is a reductive (inverse) proof that the amions have a normative velocity. Here, I use the so-called **method of reductive analysis,** or backward analysis.

First, we create a general composition of all sciences (the theory of everything); then the achieved combination (the congruent correspondence of the sections of physics) becomes a complex proof of our accepted postulate for the creation of a new evolutionary theory. In particular, we have determined that the property of inertial motion and the amion aser have a normative constant value and correspond to the mass of the particle. The property of inertia has both a quantitative value and an original form: the amions tend to rectilinear translational motion. Thus, we find a new complex proof that the amions have normative speed and aser.

'Entering the society' and creating a structure, amions formerly living on their own – in the chaos and lawlessness – are already forced to obey the public (structural) interests. Society as a dynamic vortex structure requires the amions to accelerate the movement it needs, and their resistance to violence gives rise to Newton's reactive forces, which are called the Coriolis force. The system makes the amions walk in circles, which is unusual for them; their corresponding resistance leads to the appearance of the centrifugal force in the general system.

Now, I'd like to explain the reason for the increase in the speed (acceleration) of air molecules in the compaction of the amionosphere (climate issues). First, we'll return to our car for a minute. It moves by inertia despite the adhesion of the wheels to the roadbed, while the deformation of the wheels and roadbed damps part of the inertia of motion. To compensate for the decreasing speed of the car, the engine sends portions of the circular motion to the wheels, and then their grip gives this motion an additional impulse. I note that the grip of the wheels also depends on the quality of the roadbed.

The movement of air molecules occurs in a similar way: their 'wheels' cling to the amionosphere. At the same time, planets, molecules, and the car are moving forward due to internal inertia forces. Molecules roll by inertia, like a thrown ball, and also receive an increase in speed as a thrown and simultaneously spinning ball. The rotation of planets and air molecules around their axis gives an increase in the forward inertial motion; it's as if they are drilled into the amionosphere. The self-rotation of planets is analogous to the self-rotation of atoms and molecules.

In the same way, I explain the motion of gas molecules, answering the question: why do gas molecules accelerate when the air temperature rises? Today, physicists explain this phenomenon the other way round: the effect of electromagnetic waves accelerates the molecules, and the increase in the kinetic energy of the molecules gives an increase in heat in the atmosphere. I, for one, came to the conclusion that compaction of the amionosphere leads to an acceleration of the gas molecules – the warming, which is the aggregate (sum) of the kinetics of amions **mv**: the more amions per unit volume of the atmosphere, the higher the air temperature.

To organize good adhesion to the amionosphere, each planet and molecule has its own amion cloud (rubber tires). This idea is proved by the acceleration of air molecules as a result of amionosphere compaction. Perhaps, you will say that this is sophistry. But I will argue that this is a deduction based on reductive analysis. We know that as the temperature of the air increases, the speed of the molecules increases, and this is perfectly explained by the adhesion of rotating molecules in the amionosphere.

The processes taking place in the Solar System, namely, the initial turbulence of particles in space and the accelerated rotation of a large cosmic sphere, the centrifugal force, as well as the Coriolis force, do not occur sequentially – they occur practically parallel and permanent in the form of microtendencies. They induce each other by impulsive microscopic influences. Microtendencies are one of the main ways of evolution: all changes in the cosmic sphere occur with microscopic impulses; at the same time, different types of changes are mutually supported.

Fifth Stage of the Physical Evolution of the World

In the center of the ellipsoid, the predominant part of the amionic matter has already been concentrated and transformed into a protostar (the future Sun), and local vortexes in different concentric circles have formed into protoplanets (see Figure 23).

The properties of attraction of the amions (inside) lead to spontaneous compression or compaction of a large system and its subsystems – protostars and protoplanets. Local vortexes have been snowballing, collecting most of the free amions inside the ellipsoid of a large system (a cosmic body) in the protobody of planets.

Noting an important matter: when the mass of the star and planets increases, the depletion and rarefaction of the interplanetary medium simultaneously occur. The planets begin to have their own satellites, and this is third-level vortexes. The first level is the entire Big Atom (the Solar System), the second level is the planets, and the third level of vortexes is the formation of planet satellites.

After all these transformations, the Solar System changes from the primary evolutionary medium into the evolutionary system.

Gradually, a low-entropy amionosphere is formed between the planets, which means the birth of an inner cold space. At first, the Big Atom had a gradual transition of entropy, or density of amions, which decreased from the center to the periphery in a quadratic function. Now, in the general system, new formations are generated, each of which has its own center and its own diminishing periphery.

Previously, we stipulated that the complete coordinated movement of the planets and the central star is no longer there, so the protoplanets and their satellites slip in the general amionosphere of the Sun. At the same time, like a snowball, they literally 'sweep' it around the entire concentric circle. This process clearly illustrates the modern state of Saturn, the rings of which can be seen as a result of belated self-formation of this planet.

In the photographs taken by the Voyagers (American spacecraft launched in 1977, which was the first to reach the boundaries of the Solar System), one can see small satellites in the empty gaps between the rings of Saturn. It can be regarded as a product of the initial stage of formation of the entire Solar System. This planet has not yet compacted into a plasma state; the Coriolis force are somewhat weakened on the periphery of the Solar System. It continues to purify and lean out the adjacent space, which may not end if the Coriolis force has exhausted itself there.

On the example of Saturn, we see the processes of formation of the planet satellites. Let us also note the following: distant planets are gaseous because they have not completely emerged from the state of primary space formation.

The density of the amionic matter in the center of the Solar System reaches its maximum value, so the protostar has a shape close to the sphere. In the amions of the central part of the Solar System, part of the kinetics is transferred to amions on the periphery, so they have predominant aseric properties; it can be said that they are positively charged. At the same time, the amount of kinetics in the amions on the periphery has grown; therefore, they have a negative charge. See Figure 13.

In the central part of the large system, the speed of each individual amion decreases, and the aseric properties cannot change, so there is much more aser in relation to the kinetics (mv). Therefore, according to Avogadro's law, the density of amions increases and the central part of the Solar System is shaped like a sphere. I will explain with an example. Mercury takes the form close to a sphere, and the droplets of water on the leaves have the shape of a crushed ball because mercury atoms have slightly more aseric properties than kinetics, and the water molecules have slightly less aseric properties. At the same time, both substances are suppressed by the forces of Earth gravity. To recall, we said that stretching the ball into an ellipsoid is due to the predominance of the amion speed over the aser properties, and at the center of the system, the predominance of the aser leads due to the fact that the system is shaped like a sphere. Here, both the reduction and deduction of the analytical method from logic are evident. We can check the solution of the problem and its answer through the imposed statement of the problem, and we can also verify the statement of the problem (postulates) through the answer and the solution.

As it moves away from the center to the outer boundary of the ellipsoid (the crust of the Big Atom), the entropy (density) of the amionosphere decreases in a quadratic function. The whole aggregate of particles in the interplanetary space (amionosphere), as well as the planets are connected by a voluminous network of forces of attraction. At each point of the solar system, there is a dynamic balance between the forces of attraction and the forces of inertia of the particles. So in its amionosphere, there is a balance between the proton center and its electron planets. The protostar (the Sun), the planets, and the amionosphere (cold space) constitute an organically integral structure. We can say that together they make up the Big Atom.

The bulk of the amions in the large system is accumulated in the center, which continues to rotate the entire amount of amions within the system. The center sets all local formations in general motion. The planets move not only by inertia; the center 'helps' them through the redistribution of forces along the chain of amions.

The rarefied amionosphere (inner space) continues to serve as a transmission mechanism; it organizes the coordinated rotation of the star and planets with its volumetric network of forces of attraction.

It is noteworthy that the rarefaction or depletion of the amionosphere in the interplanetary space stops when it comes to 'servicing' the entire gravitational system (no longer an aseric network) of the Big Atom. The remains of amions in the interplanetary medium occupy 'regular posts.' They serve the general system of gravitational forces emanating from the center. The common volumetric gravity network (a network of gravity forces from the center) retains amions in the structure of the Solar System so that the planets cannot further deplete the interplanetary medium.

The sun sends its powerful forces of attraction and colossal kinetic energy of rotation to the planets through the gravitational bonds of the regular amionic structure – the amionosphere. The entire large amionosphere is an organic continuation of the central star, so it rotates with it and thereby carries all planets away in the general dynamics of rotation. Continuing to act, the centrifugal and the Coriolis force rotate the planet around its axis, but in the opposite direction. Let's recall the example of a wooden bar floating in a box with water. When the cart moves along the trajectory curvature, the bar makes a triple move; in this case, our planet is a bar and the water is a cold space. We said that if the liquid in the box was viscous, it would slow the movement of the bar, i.e., the water/space affected the movement of the planet/bar.

However, the compaction of amions (particles) in the center and on planets stops at a certain stage of formation of the dynamic equilibrium of the Solar System. Dynamic equilibrium in a large structure is the balance of two conflicting forces, such as kinetics at the periphery and the forces of attraction emanating from the central part.

On the one hand, this is a simple sum of the inertial energy of the entire aggregate of amions in the structure; on the other hand, this is countered by the integrated sum of the attractive forces of amions of the periphery and the entire system. At the same time, there is a voltage in the gap between the center and the crust of the atom, which is called system springing. This phenomenon is formally expressed in the existence of the ecliptic of the Solar System.

The star and planets stop condensing at certain stages. At a certain critical moment, the center of the cosmic body structure ceases to compact the protomatter with its organizing forces (the aser properties of amions). The simple geometric tightness does not allow further compaction of the Solar System. An example of this is Avogadro's law, according to which the geometric narrowness of self-moving (restless) molecules does not allow them to compact in a limited space.

Before reaching the limiting density state of the center, the inner (future stony) planets have already compacted to the plasma state.

At the same time, now we cannot make an unambiguous conclusion about whether the self-compacting of the gas planets has stopped or not.

In any material structure, the forces of attraction slightly dominate the kinetic energy of the entire aggregate of amions. Otherwise, it will be unstable and inevitably disintegrate. For its existence, the organizing forces of attraction of amions must exceed their aggregate kinetic energy. In other words, beyond the material structure, residual forces of attraction always act, which come into balance with the entropy of the external medium.

In this chapter, and for the first time in physics, I use the concept of atomic cortex. The difference in density of the amionosphere of the Solar System and the galactic space leads to the formation of the crust of the entire Solar System.

The crust is a compacted boundary of any complex material structure, without which the existence of a structure, its dynamic system, is impossible. In addition, certain structures also have a virtual boundary.

Any material structure necessarily aligns its descending entropy with a poorer medium. A leap in the chart of the Big Atom's entropy takes place in the atom's crust. See Fig. 28. But outside the crust of a complex structure, there always remains an excess of gravity forces. Passing through the crust, the forces of attraction of bodies and the Big Atom decrease and automatically equalize with the entropy of the external medium. In the physical sense, if there were no crust, then there could not be substances and bodies. The absolute value of the forces of attraction outside the structure crust must always correspond to the entropy (density) of the external medium.

On the contrary, if a physical body such as a salt crystal is placed in a molecularly tight discrete medium with high entropy (high density), for example, into water, the crust will immediately begin to dissolve. The effect of the [magic] phenomenon is that the body (substances) in a rich medium no longer needs crust – it becomes physically superfluous. After the dissimilation of the crust, the body dissolves in a dense medium up to individual atoms or molecules. In this way, even an ingot of gold can be dissolved in a mixture of two acids. The crust as a part of the body shape is the forming element and the keeper of the complex structure.

Fixing our tradition to briefly leave physics for the humanitarian field, let's draw some parallels. First, people consider themselves and the things that are already familiar to them as the closest of all. This method is sometimes called parsimony – the use of the simplest possible explanation. Secondly, it contributes to the formation of evolutionary logic and a universal way of thinking.

As an analogy of the physical concept of the crust that we have introduced, we can mention the Great Wall of China. Rich countries always strive to strengthen their border with their poor neighbors, and this is no coincidence; the laws of physics in a sublimated form are repeated in human history. The crust of any structure is always on the border with poorness. If, suddenly, an unidentified mini-state with a natural economy, a plow, and horse transport is discovered on the territory of the United States, then the border will be of no use to it.

For uniformly rich countries of Western Europe, the border became a formality as the idea of the European Union was implemented. However, the problem of borders in the European Union became relevant again when illegal migrants – people from poor countries – began to enter its territory.

Lands with fertile soil and bowels of the planet represent value for people; we need them to meet our increasing needs. But we suddenly start to consider protection of the climate, the atmosphere, and the pristine nature as our main value or, let us assume, the whole world will be united by the general idea of salvation from any imminent threat (such as the attack by aliens from outer space, as a fantastical example), then the problems of borders between countries will recede into the background. On the contrary, the task of creating a world government will appear. The laws of the universe (physics) act through the whole complexity of the world. Therefore, we, the humanity, will sooner or later become aware of the correct non-violent adaptation in the biosphere.

To solve modern problems, only strong ideas can unite people around the world; the idea should have the penetrative power of the **lightning leader.** Knowledge and science become the driving force of the era. Like any force, it has a direction – a vector of attention and scientific research. The task of this book is to form the organizational beginning of science against active forces, which then direct it to the solution of pressing problems.

What motivates us? We are motivated by the vital energy given by birth; it is the sublimated simple inertial force of a multitude of amions in our body. For example, children continuously move and do something. But if their activities are not channeled in the right direction, they will destroy everything that will fall into their hands, because they are not yet trained to create; they are guided by a simple motor energy not burdened by the organizational beginning of consciousness. They can even set fire to their own house and die. Children's activity is simple; they do not yet have knowledge and awareness. People on Earth are like children; they will remain children until they master the new ideas of evolutionary logic.

The material crust is necessary as a formal boundary for the existence of complex structures in order to maintain their density. The structure cannot exist in the void. In an absolute vacuum, any atom will melt or burst like a rubber ball without atmospheric pressure. The structure must rely on the entropy of the medium and adapt to it. This definition implies the need for consistency between the complex content and the external form.

Let us consider the history of occurrence of the Solar System's first crust. First, there was an accumulation of amions in a quiet corner of space without pronounced shape and boundaries. We have already decided that this is the so-called dark matter. The center of gravity arbitrarily formed in this cluster of particles. Then the cluster took the form of a sphere that absorbs light, but does not yet shine. It is at this point that its crust is formed. At the boundary of the large sphere and outer galactic space, there is a difference in the entropy of the internal and external medium.

If we compare the Solar System with a cosmic ball, then this ball must build its own 'rubber crust' in order to maintain internal pressure and high entropy. The medium, or the outer space, has a stable but very low level of entropy (density), and the Solar System must strike an entropic equilibrium with the space as a medium. For this purpose, it builds a dam – a crust.

The density of amions within the future Solar System is very high, and the entropy of outer space is much lower. There is a large difference between the density of the amionosphere of the large sphere and the low entropy of the outer intergalactic space. The large sphere behaves in the only possible way – it creates a material crust on the border with outer space, building the conventional Great Wall of China. The equilibrium of stresses must be ensured and the different levels of entropy (density) must be balanced.

Figure 28 shows the chart of entropy (amionic density) of the Large Atom. Amions in space are connected by forces of attraction along the X, Y, and Z axes; each axis has two directions. We see that we are talking about a three-dimensional space, where the amions have six bonds with the other amions. In reality, each amion has three personal aseric links. The other three of the visible six bonds belong to neighboring amions. Thus, each amion has a need to alleviate three aseric bonds in space.

Imagine a ball in the form of a multilayer sphere, like a bulb onion. Let's suppose there is a certain spherical surface in its very center, where a lot of amions are located; their number depends on the surface area (we take Avogadro's law into account). Then there is the next layer, where the number of amions will be the same. It is clear that as the layering of surfaces increases, the total area of the subsequent sphere also increases.

Why did we accept the condition that the number of amions in the different layers of the 'bulbous' sphere will remain constant? This is due to the principle of 'three links,' which has an amion in the first layer, including two in its own layer, and another in the next layer. The thickening of the crust becomes possible because of an excessive unclosed bond in the last layer of the sphere. The unfulfilled last bond attracts one more amion from the external medium, which needs two more connections and so on. Figure 29 shows fragment of an atom in axonometry; we see the empty center and the crust of the atom. We see that the inner layers of the atom should have the same number of amions in each layer, but the crust needs more amions.

It will be appropriate to clarify Newton's law of gravitation – the inverse dependence of forces on the distance between bodies. The density of amions in the peripheral layers of the atom decreases in a quadratic function, which will later be repeated in the gravitational network of the planets (cloud). See Fig. 29. Figure 30 clearly demonstrates that the force vectors, or the aseric chains of amions, are structured as rays emanating from the center of gravity of the bodies. We see that the density of amions decreases and the gravitational forces are inversely related to the distance between space and physical bodies.

Outside the Big Atom, the amionosphere has very low entropy. The feature of the crust is that it is built by a newly emerging structure. As it thickens, this structure also increases the crust density. When the water begins to flow into the reservoir, we try to strengthen and build up the dam to keep the growing volume of water within the necessary limits and avoid flooding. In our case, the pressure builds up from the inside and the body of water gradually self-compacts; respectively, the external environment becomes relatively poorer, and thus the crust becomes denser.

The first crust appeared on the second stage of the Solar System formation described in this chapter. Then it shrank in size along with the sphere and simultaneously consolidated and deformed as the volume of space formation decreased. Today, the crust of the Solar System is probably located behind the Kuiper belt, where the forces of gravity align with the very low entropy of outer galactic space. See Fig. 28 and 30.

The phenomenon of crust formation is known to us from our childhood: we observed it in the form of skin on the surface of hot milk. The physical meaning of its formation is that the autonomous structure of substances (milk, liquid) is balanced with the external poor medium – the atmosphere. The liquid aggregate material is leveled with the entropy of the external medium, building a dam.

In classical physics, the appearance of a film is regarded as the emergence of surface tension forces. The physical meaning of the film as the atom or substance structure's needs to strike an equilibrium with the external environment is not considered at all. It is because scientists believe that atoms are in a vacuum. In reality, atoms are always in the medium of amionic matter; this rule is common for gases, liquids, and solid matter. A complex structure with a crust becomes an autonomous structure, simultaneously entering into an internal and external thermodynamic equilibrium. The crust plays the role of the body of the vessel, inside of which is the 'rich' content of the structure, which tries to preserve this wealth and, therefore, continuously strives for entropic equilibrium with the medium.

The Sun cannot have a crust because it is only the core of the Big Atom; the star is not an independent thermodynamic system. On the contrary, all planets, as well as other physical bodies, have their own crust and their own autonomous dedicated amionosphere. The Earth crust has its borders and rotates with the planet, so artificial satellites are launched in the direction of rotation of the Earth and its amionosphere. In the opposite direction, the satellites cannot be in free inertial flight. Synchronization of the satellite's flight with the rotation of the planet is called the finding of a satellite by means of its geostationary orbit.

I'd like to once again compare the crust of the Big Atom with the dam: it is built on rivers to keep the water pressure level on the upper water head in a certain accordance with the level of water pressure on the downstream. The last or the lowest water head can be roughly compared with the magnitude of the forces of gravity. The crust needs an atom to hold the density inside the structure of the atom. Through the crust, the atom aligns its internal density with the density of the external medium.

After detailing and mastering the concept of the crust, we must return to the question of the original forces. As follows from the foregoing arguments about the crust, most of the kinetic energy of the amions is blocked by the forces of their attraction; only a small part of these forces are directed to the external medium. See Fig. 28 and 31.

With the help of residual inertial forces, complex structures move in space, and residual nuclear properties (gravitational forces) act as forces of gravity. At the same time, the assumption that we see only the blocking of the forces of attraction, and the inertial and kinetic power of the planets continue to act, is fair, since the kinetic forces of the planet can only affect the environment through the forces of gravity. If you have a huge train passing by near you, it will be soundless; most likely, you will not even notice it. But if it has pronounced original forces of attraction, they will entice you.

The introduction of the notion of primordial forces, or rather, the properties of aser and inertia, is important for the implementation of a radical change in the understanding of the laws of fundamental physics in the context of evolutionary logic. We usually do not see nor feel the primordial power of the basic properties of amions in nature. People saw the surrounding world already in a civilized, orderly, and 'tamed' form. As a result of billions of years of nature evolution, its self-development, the original chaos of space has transformed into a modern complex but orderly world. The a priori properties of amions are hidden inside the atoms, under the crust; we first felt their primordial power only after the explosion of the atomic bomb.

At the same time, based on scientific experiments, we know how much chaotic energy is 'packed' in the atoms of substances. Billions of years ago, our planet was in a state of plasma; the temperature on its surface reached several thousand degrees. As a result of the evolution of the physical world and the complication of matter, the kinetic inertial energy of all amions is blocked in atoms by the aser property, or by the forces of attraction of the amions themselves.

Today, we live in a world that is 'tamed' through self-complication of matter. It is with this understanding of the new foundations of physics that a new **evolutionary theory of climate**, outlined in the next chapter, begins. In the modern world, there is only a residual manifestation of wild primordial energy. In fact, Einstein's formula for the total energy $E=mc^2$ somewhat distorts the overall picture. A huge mass of matter is hidden in the atom. The mass manifested depends on the real mass of the atom; there should be an average transition coefficient.

In the Einstein formula, the nominal velocity of the amions inside the atoms (substances) is incorrectly accepted as the speed of light – at its maximum possible value. I think that he did it intuitively, without understanding the structure of the atom. In this formula, the mass must have a much larger value; respectively, the normative speed of the particles must be greatly reduced. We can say that by this formula, Einstein proves his own words about the random character of the formulas (see Chapter 1).

At the very beginning of the book in the postulate of the new beginnings of physics, we accepted the thesis that the amions have normative inertia of motion and a standard speed, which should be much less than the speed of light. It can be calculated through the orbital speed of planets and the speed of other observable objects in space. In this book, we will not delve into this and other questions about specific numerical parameters. The main task is to develop new physics beginnings for understanding climatic processes.

Amions reach a high (limiting) speed when they are discharged by micro-explosions (nanovortexes) from the solar plasma. At the same time, they form a very high frequency helical wave motion in the amionosphere and organize a stream of sun rays that reach the maximum possible speed – the speed of light. See Fig. 26. The figure shows how nanovortexes fly out of the solar disk in the form of rays, and the released space is occupied by amions from the close environment.

They continuously replace the place of ray vortexes under the influence of powerful forces of attraction of the plasma center of the Sun.

In a sense, the Sun is an eternal engine; there are multi-level processes of Coriolis vortexes in its depths. In the bowels of the Sun, there is an intense struggle between the forces of attraction and kinetic forces; they are the ones that give rise to hundreds of levels of countless vortexes. It is only occasionally and in extreme cases that we observe the manifestation of the primacy of forces. For example, as already mentioned, we actually saw the intra-atomic kinetic energy after a nuclear explosion. We see the primordial nature of the forces of attraction and the kinetic energy of amions in electrodynamics: invisible electrical forces manifest themselves much stronger than gravitational ones.

Electric forces arise under a certain order of interaction of two simple forces in a very complex structure. The atom is like a compressed spring – the forces of attraction inside the atom 'squeeze' its kinetic energy. Sometimes, under certain conditions, this spring is slightly compressed and unclenched, and this is a real phenomenon of electricity.

We came to the conclusion that at the beginning of time, at the first stages of the physical evolution of the world, in the formation of the cosmic sphere, the processes of densification and vortex in the protomaterial world occurred intensely because of the primacy of the forces. They were able to compact the protomatter to the plasma state of the center of the Big Atom in the form of a protostar.

Modern physics has come to understand that atoms rotate around their axis; this phenomenon is called spin. If we take the structure of an atom identical to the principle of constructing the Solar System, then electrons in an ordinary atom should appear under the influence of the Coriolis force; at the same time, they will be somewhat unstable due to the small size of the atomic structure: they will randomly appear and disappear. The electrons in the atom are the vortex of the amions at the periphery of the atom, like the planets in our solar system, and the protons are the central compacted part of the atomic structure. The proton is similar to our Sun. We can assume that the Big Atom, such as the Solar System, and a micro-atom in the microcosm are identical in construction principle; both have a multilevel vortex design.

The Big Atom differs only in size, so it was in a large system that stable stony planets could form. The microatom has a compacted center; it retains the totality of all the amions in the atom structure, while the attractive forces of the center retain the kinetic (inertial) energy of the amions. The entire structure of the atom rotates around the axis of symmetry, and the kinetic energy of the amions is redirected to rotation around this axis.

The nucleus of the atom also rotates; it rotates around the emptiness, the 'eye' of the natural vortex that creates the amions, making a concerted movement. The density of the atom is controlled by the forces of attraction of the amions. The micro-atom must have a disk-like shape stretched out in the equator since the Big Atom has a pronounced ellipsoid shape. On the surfaces of a spherical and ellipsoidal shape, both atoms form a crust – a dense structure of amions.

When creating the atom, nature took the only possible path. The amions continuously move and simultaneously attract each other. The a priori properties of the amions make it possible to create a complex structure with a certain form of motion. The vortex structure of the atom is the only possible way of organizing amions in the form of complex structures. Such form of motion is created by a combination of their properties.

All areas of physics, including thermophysics, nuclear physics, and electrodynamics, return to the mechanistic beginning. We have created a new version of understanding the processes in the microcosm based on the impulses of reactive processes (the Coriolis force + the centrifugal force) and the phenomenon of implosion. Does the uncertainty principle of quantum physics have any meaning for us? Answer: yes, it has, since these contradictions become irritants; trying to get rational answers to them, we find new solutions.

At the same time, we see a fundamental difference between the laws of physics and the laws of living nature. Before the advent of biological forms, the laws of nature that govern the physical structure are tough, resolute, and unambiguous. In physics, Laplace's principles of cause-and-effect determination are observed. It was only after the appearance of living forms that ambiguity appeared.

The Solar System is a great cosmic mechanism; its direct purpose is to serve as the cradle of evolution of our planet, including the appearance of humanity and its specific development. For this purpose, the Big Atom creates the 'full range of services': rhythmic variability of the medium, a source of energy in the form of the Sun, and cold space. The 'clockwork' of the planetary system is formed; it supports the unique rhythm of the medium variability. The orbit of our planet has a small eccentricity and is almost circular, it slightly 'rocks.'

Our planet rotates in the right cycle of time and meets all the requirements; therefore, it was chosen as the object of biological evolution. The Earth is continuously exposed to the sun rays, but half of the planet is also being constantly cooled down in the inner space. More precisely, the planet is not cooled down, and the irradiation process is interrupted; at this time, diffusion processes and a new structuring of the amionosphere in height occur. When rotating around its axis, the planet experiences conflicting influences in a daily rhythm. It 'relaxes' under the influence of solar energy, seeks and finds new forms; then it is cooled down by the space and fixes the solutions found.

After we adopted the binary opposition of the properties of the smallest particles – amions – and adopted the principle of 'two in one,' we crossed the Rubicon in the evolution of our consciousness and are ready for new breakthroughs in science with all complex structures, starting with the atom being viewed as a vortex complex structure with a center and periphery. At all levels of complexity, we will use the universal analysis method based on new evolutionary logic.

The scientific model of the world reaches perfection and begins to explain many of our life's complexities in the blink of an eye. A shift takes place in science, in the worldview, and in the public consciousness, and each of us enters a new level of thinking and adaptability. We acquire the mind of expansive and panoramic action as a tool of navigation in this rapidly changing world. As a result, a new and perfect person is born on the 'superposition of science,' who already knows that the whole planet is an integral evolutionary system, a single organism, and it is the modern person alone who is in the focus of evolution.

Chapter 3. EVOLUTIONARY THEORY OF CLIMATE

Today, problems related to climate change cause concern to the entire world community. They affect every single one of us, raising questions. In the second half of the twentieth century, scientists of developed countries undertook to solve new problems systematically and thoroughly, engaging relevant resources. Ecology and climate constitute a challenge of the planetary nature for modern science.

As a rule, the impetus for research is the observation and measurement of the surrounding world and the accumulation of factual material on the problem of interest. Based on the results obtained, scientific theories are created. The global goal of all theories and science, in general, is to anticipate future events, ensure the safety of people and the environment and lay the foundation for a sustainable development strategy for each country and the world.

I believe that the creation of a separate scientific theory of climate in isolation from the **general theory of evolution** is impossible because the climate is only an indicator of the quality of the evolutionary system, including the habitat. In our era, the quality of the environment and the pristine nature depend (and will depend even more in the future) on the activities of people, the general level of development of science, and social self-awareness. The world science will have to solve (or not solve) the grandiose tasks of unprecedented scale, which for various reasons have been put on the back burner.

The order of things, which has established on Earth over millions of years of evolution, and nature as a whole counteract human influence and our attempts to redesign them quickly; quickly by the standards of evolution. Under the existing approach, all our endeavors can result in inevitable failure. We are increasingly convinced that we need a full scientific approach to relations with nature; we must reliably know what we can and cannot do.

The new theory of climate is created as a follow-up to the basic ideas of the general theory of evolution. As a separate area, it acquired the name **Evolutionary Theory of Climate (ETC)**. The planet's climate is the main subject of research in The Canons of Evolution; its object is the entire planet Earth as part of the Solar System.

The formation of the climate at the stage of physical evolution will be considered in the first sections. According to the logic of evolutionary development, this part of the chapter can be called "The Sixth Stage of the Physical Evolution of the World" and move to the end of the chapter, "The Formation of the Solar System," where we reviewed the previous five stages.

At the sixth stage, the structural content of the climate is formed on the basis of the **great ratio of land and the world ocean** – the ratio of their areas and physical qualities. I call it great because it lays the foundation of the climatic mechanism and creates the prerequisites for biological evolution. The world ocean and then the land were the medium for the emergence and development of life on the planet. In this chapter, we will see the transition of a leading role in the formation and mitigation of the climate from inanimate forms of matter existence to living biological forms.

To recall, in the chapter, "The Formation of the Solar System," we examined the compaction of the central part of the Solar System to the state of plasma and the formation of a protostar and then saw the formation of planets and their satellites and learned about the beginning of the processes of implosion (inwards explosion) in the body of the protostar. In physics, the phenomenon of implosion has not been previously considered from such a perspective as a star being lit up.

As a result of implosion processes, the solar plasma began to emit electromagnetic waves in various ranges, after which all planets and their satellites were exposed to solar radiation (irradiation). Considering that the planets rotate around their axis, the daily variability of the medium appeared on their surface, which we will discuss in detail in the relevant sections. It gave impetus to the formation of the climate and its evolutionary development on the planet Earth.

Of all the latest advances in climate science, I highlight the discovery of **epochal cyclicity**, because I consider it a fundamental factor in evolutionary development. The epochal cyclicity of the climate appears at the biological stage of development. Its interpretation is ambiguous and has caused controversy among scientists. The prevailing view is that epochal cyclicity is the result of external influences, primarily the strengthening of solar activity. The fundamental difference of the evolutionary theory of climate is that the internal processes taking place on the planet are considered the cause of epochal cyclicity. A special section at the end of the chapter will be dedicated to this issue.

Summarizing the introduction to the chapter on the ETC, I want to emphasize that because of the contradictions in scientific theories, we still do not have an answer to the main question: **Will humanity find the real causes of unnaturally rapid climate change and be able to correct the situation?**

In the first two chapters, we laid the postulate for new physics beginnings and built the hypothesis on the formation of the Solar System on its basis. Now let's proceed to the solution of the main task of The Canons of Evolution and start developing the ETC, or the Evolutionary Theory of Climate.

What is Climate? Some Clarifications

Climate is the result of physical, and then simultaneously biological evolution of nature over billions of years of the planet's history. In the Evolutionary Theory of Climate, climate is the physical state of the environment on the planet surface and in the lower layers of the atmosphere. Its main characteristics are temperature, humidity, and pressure. The climate is variable during the day, which is due to the impact of sunlight on the surface of the Earth and the rotation of the planet around its axis.

Evolution takes place in the form of microtendencies in animate and inanimate nature, a gradual complication and development of the first physical structures, including atoms, substances, and the crust of the Earth, and later in the form of biological structures. All this together led to the formation of a stable and mild climate.

By **climate stabilization**, I mean the gradual decrease in the average temperature of the planet over a year, as well as a reduction of temperature fluctuations in the lower layers of the atmosphere.

A favorable climate is the predictability of weather conditions, a reduction in the frequency of natural disasters, and a decrease in the top-to-top weather values. The progressive decrease in the frequency and force of extreme climate fluctuations during the long evolution process provided the conditions for the corresponding progressive complication of matter and the development of complex, biologically productive plant and animal species. In turn, they provided reverse moderating effect.

For example, the mitigation of global climate created conditions for the development of tropical forests, which then provided mild favorable conditions for biological diversity. As a result, all the necessary prerequisites for distinguishing a person from the animal world appeared.

Evolutionary processes are a complex of very sophisticated subtle components, by which only biological evolution is usually meant. I call the complex of physical complication of primal matter and the formation of the solar system **the physical evolution of the world**, which gives rise to **biological evolution** at the final stage. The evolution of biological forms is a natural process of nature complication; this is the next stage of the general evolution and the continuation of the physical evolution of the world. Then, on the integrated basis of the two stages, the evolution of person and human society begins.

Basic thesis of the Evolutionary Theory of Climate No. 1. *Climate is the achievement of world evolution. The modern soft and stable climate was created* *by nature over billions of years of physical, and then biological evolution. Gradual mitigation of climate is a necessary condition for evolutionary progress.*

Modern science cannot unequivocally and reliably explain the main causes of climate instability. There are three main theoretical movements, although not one of them has yet been supported by most scientists.

A significant part of them considers the destabilization of climate a consequence of the spontaneous development of humanity, believing that climate change occurs due to human fault and is a consequence of structural disturbances in nature. They refer to the destruction of half of the land cover and the violation of the mechanism of climate self-regulation. Their main thesis is that it is the man who violates the delicate balance achieved by evolution in nature. But this thesis is of a general nature; there is no rational scientific theory in its basis. To become a guide to action, it needs a reliable scientific justification.

Another group of scientists considers the influence of people on nature and climate as an insignificant factor. They see an increase in solar activity or other reasons as the main cause of climate change and recognize the climate destabilization part of the natural epochal cycle. Their conclusions look quite scientific in their form and are based on seemingly scientific theories and calculations. At the same time, they cannot in any way explain the accelerated rates of the processes of aridization and the degradation of nature. For example, they do not have an answer to the question of why the epochal climatic changes (rise and fall of average temperatures) occurred hundreds of times slower than now. Back then, it took about 10 thousand years to increase the average annual temperature by 1 °C, while now this happened in 100 years, and there is a tendency for acceleration.

There are also skeptical scientists, but there is not many of them. They consider the long-term climate prediction to be mathematically indeterminate due to the multifactory and complexity of climate processes.

The main and fundamental difference of the proposed new theory of climate is its main idea, which is contained in the title. This theory of climate is evolutionary. In this chapter, I articulate seven basic theses of the Evolutionary Theory of Climate, which form the foundation of a new theory and are based on the new beginnings of physics and the postulates outlined in the first chapter of the book.

Modern climatologists and physicists adhere to a narrow interpretation of the law of conservation of energy and the three principles of thermodynamics; they view our planet as a passive cosmic object. In simplified form, it looks like this: during the day, the planet is heated by solar radiation, and at night, it cools down in cold space, just as the smith first heats the metal blank on fire, and then cools it down in cold water.

At the same time, physicists do not take into account that the planet is an organized complex structure, and the near space, including the Moon, is merely a continuation of the body (organism) of the planet. In the terminology of modern atomic theory, the Earth is a proton, and the Moon is an electron; all dynamic processes occur within a single atomic structure.

How does the planet warm up and cool down? During daytime irradiation, the surface of the planet emits heat to the lower layers of the atmosphere, and at night, it assimilates (or absorbs) the previously released heat back. This thesis is fundamentally inconsistent with the generally accepted position that the planet is cooled in space.

Our planet is a 'cosmic atom'. It has adaptive properties due to being part of the Big Atom – the Solar System. The planetary atom is only structured within its boundaries, which are beyond its satellite vortex, the Moon. We will consider this process, or more precisely, the complex of processes, in detail in further sections of this chapter and conclude that the planet cannot exchange thermal energy with space; it only passes from one entropic state to another and can generally return to the hot stage (God forbid).

The main idea of the Evolutionary Theory of Climate is that the climate is created and stabilized by new structures on the surface of the planet (the great ratio of land and the world's oceans) and is further mitigated by the development of the biosphere and cryosphere. A dialectical relationship arises: microtendencies of climate mitigation contribute to the emergence of new complex biological species, which, in turn, leads to further climate mitigation and, accordingly, causes another wave of complexity and development of wildlife.

Winter House: The Climate Model on the Window Glass

We will conduct a climate experiment in a detached little house. Let's say its window has single glazing. What will we see on the window in winter? In the middle of the glass, there will be a dry and transparent part. Along its perimeter, we will see a fogged-up area, and on the edges, there will be a frost, icy patterns.

If we purposefully enhance the action of the heating appliances, then, as a result of increase in temperature inside the house, the situation on the window will change very quickly; the area of the dry section will increase, the fogged-up area will remain only at the edges, and the ice will disappear altogether.

The ratio of different zones on the glass resembles the state of all the land on our planet. The dry space of the window is a desert; frost around the edges are polar glaciers; the condensed area is the remaining, humidified and biologically productive part of the land. The dimensions of the areas of different sections of the glass depend on the ratio of the indoor and outdoor temperatures. This glass, as a monitor, reflects the tremulous entropy balance of the microclimate of our house and the outside temperature.

In the dynamics, the cold outside air and warm room air interact through the thermal-insulating properties of the walls. The climatic experiment in the winter house and the state of the planet climate differ in that, with the help of the glass, we are saved from the cold; wildlife, on the contrary, is mainly concerned with protection from the sun's rays. Both options share a certain materialized boundary – a layer that protects them from external influences. The house has a boundary in the form of a wall fence, and our planet has its green cover and clouds. In the experiment, the solar energy is replaced by a heating system, and the increase of heating appliances in the room means that we have increased the area of exposed land and enhanced the reactive quality of land.

Thus, on the window glass, we saw the climate action of the planet in miniature. The only difference is the scale of the planet, the entropy inertia of the Earth's mass, and the presence of an aquatic medium.

Two Levels of Climate

I have determined the two levels of climate conditionally. In reality, they are inseparable. I define the first level of climate as a result of the physical evolution of the planet and call it basic. The second level is the result of biological evolution; I called it gentle.

The first level is the basis of the second; it is a symbiosis of two worlds. The sun and water destroy the rocks, the wind scatters sands all over the world, and the living nature reclaims sand deposits and creates a humus layer – the soil. The processes of mutual influence of the two worlds have been continuously going on for hundreds of millions of years. And it is only humans that managed to weaken and bring these processes of evolution to a critical level.

Basic thesis of the Evolutionary Theory of Climate No. 2. *There are two levels of climate: basic level, which is a result of physical evolution and gentle level, a result of biological evolution.*

In other words, there is **basic climate** and **gentle climate**. The first, or basic, level of climate is the material foundation created by the self-development of the planet. The great ratio of land and world's oceans determines the fundamental stability of the basic climate. It is formed by a double content of the planet structure and fixed materially by solid, hot, and reactive land on the one hand, and mobile (flowing) and cool water elements on the other. They are a binary opposition; the ratio of their qualities is in a dynamic (pendulum) equilibrium.

The dense structure of land reacts sharply and strongly to the effect of solar radiation, simultaneously releasing a large amount of heat into the surface atmosphere. The air temperature in the middle part of the planet reaches 50 °C and above. The world's oceans react relatively more mildly to solar radiation; the near-water atmosphere remains cool, and the air temperature does not change sharply during the day.

The climate is heavily influenced by ocean currents; they appear as a result of inertial displacement of flowing waters. At the same time, warm currents contour the American continent from the south and the north and then mix with cold waters. As a result, they are saturated with oxygen, which is very important for the inhabitants of the seas and oceans. There is a mixing of local climates and the weather of all geographic regions.

When the planet rotates around its axis, the hot air of the continents and the cool air of the water areas are also continuously mixed, as a result of which we have an averaged and relatively low temperature of the medium. However, the main mixer of air, as well as the climate manager, are cyclones, reaching a diameter of several thousand kilometers. The climatic trouble occurred because the cyclone diameters were decreasing, and then the formation of new poles of cold began. The poles of cold are areas of the planet that are not covered by cyclones. On the contrary, the number and strength of local vortexes (small cyclones) increased in the general composition of atmospheric phenomena. Then, the duration of anticyclones declined; that is, the climate became unstable and this affected the state of the most vulnerable and tender crops, i.e. our cultivated plants.

The basic level stabilized for the first time about 500 million years ago. At the time, the climate on land was similar to that of the modern Sahara: it was extremely hot and dry, and high temperatures and dryness made biological life impossible. There is a Kazakh proverb: heat brings heat, and cold adds cold. It is etched in my memory due to its unusual logic, but now I would paraphrase it a little, simultaneously uncovering the secret of cold and hot anticyclones: the heat can stabilize the heat, and the cold keeps the cold. 500 million years ago, there were hot anticyclones and terrible vortexes (cyclones) on open spaces of the Earth, the same way it is happening on Mars today.

The gentle level of climate was created by the self-development of biological forms that originated in a favorable medium of the water area. Crawling up on land, eventually, they formed the so-called green cover of the planet, which reduces its heat exchange. The better the living nature adapted to the conditions of existence and, accordingly, the bigger the mass of the living things on land became, the softer was the climate. The softer the climate, the more complex the biological life became. This lasted until people became extremely active.

I think it is time to come up with a special name for the old/new concept, which denotes the natural green cover, or the shade clump, or more precisely, includes all existing natural formations that protect the land from sunlight and serve as a kind of insulator. In this book, I will use the word *sulde* for the new climatic and eventually evolutionary concept.

In Kazakh, sulde means an abstract or barely noticeable substantive outline of something. For example, after seeing someone many years later, you will recognize an old acquaintance by their insignificant features or voice. This is what sulde is.

In the Evolutionary Theory of Climate, this concept includes vegetation on the land surface, surface and groundwater, marshes, snow cover, glaciers, and clouds (fogs). All objects of animate and inanimate nature, which act as living and non-living conditioners in relation to the climate, together constitute the **sulde of the planet**.

When you once knew a person who was full of health, and now you meet them looking sickly, hunchbacked, and with depleted look, hardly recognizing them, the Kazakh people say: all you have left is your sulde or weak distinctive outlines. Taking sides with modern opposition, I began to understand figuratively to what state of sulde we have brought the planet. Imagine the British Isles completely covered with forests, and Western Europe too. Caesar's army literally had to chop its way with an ax, not with a sword. The first settlers – the colonists – noted the rivers of North America, saying that there was more fish in the rivers than water. For a long time, their main export product was salted fish in wooden barrels, as well as wood. Next, we will dwell on the role of nature in the climate.

For example, the tropical forest with a depth of 50 meters consistently reduces the impact of solar radiation by means of trees with broad leaves, like umbrellas, at the very top, and using plants that moisten the air deep in the forests. Inanimate nature produces ventilation throughout the atmosphere. We once again see that the two levels of the climate are inseparable. We will use the concept of sulde to denote the whole complex of natural measures for sun protection of the land.

The word sulde can be used differently: I gave the book to editors in the state of sulde, and they returned it to me in the form of a product of literature. They spent time 5 times more than their normal rate. Anastasia (editor) estimated that the volume of our postal novel exceeded the volume of the book itself. They had to translate the book into Russian but written by a person who thinks in Kazakh, because in other languages, some of our words and concepts can be represented only by an abundance of words.

The second, or gentle, level of climate will be discussed in more detail in the relevant section of this chapter. Tentatively, as a graphic picture, I would like to tell you about the ascetic way of life of tender tropical monkeys in the cold and hard environment of the highlands.

In the process of slow climatic changes, monkeys that once appeared in the tropical forests of Asia gradually adapted to live in harsh conditions of the Himalayas at an altitude of 5000 meters. In those places, they eat only lichens instead of fleshy paradise fruits, and the air at such an altitude is extremely thin. How come animals from the rainforests were able to adapt to sharp changes in their habitat?

Migrating in nature, Chinese monkeys found themselves in the region of the future Himalayas; the mountains gradually rose, the natural landscape changed, and the flora and fauna changed with it. The monkeys gradually adapted to the slow changes in the habitat and their digestive system adapted to new types of food. They found themselves in a closed system – there were mountains on all sides. In order to compensate for the rarity of air as the relief rose, the monkeys increased the volume of their lungs at microscopic rates. As a result, it grew more than twice in comparison with the ordinary species on the plain. Such changes in morphology take thousands of years.

As we can see, higher animals are able to adapt to the slow and gradual changes in climate, but they die with the rapid deterioration of the habitat, just like the entire biocenosis of the region irreversibly dies. At the same time, one needs to take into account that the plant world adapts differently; it changes the composition of the biocenosis. When conditions in the medium worsen, the structural composition of the flora and fauna changes to more primitive and less productive species.

The Physical Essence of the Climate

To find out the causes of global warming and other climate change, I try to explore the mechanisms of climate formation and self-regulation in the context of the entire history of evolution. Since the very beginning of studying the processes of climate destabilization, I have had doubts about the fundamental principles of modern physics. Many climate processes are impossible to analyze and evaluate on their basis. Therefore, my first specific scientific goal was to develop new physics in order to solve contradictions in climatology. In part, they have already been brought to your attention in this book.

Let us consider two specific positions of physics in the extended perspective.

1. In thermophysics, the concept of heat has remained unchanged for over a hundred years. In modern interpretation, the temperature of air in the atmosphere is the cumulative kinetic energy of its molecules. To solve the existing contradictions in physics and climatology, I specify that the effect of heat in the atmosphere is created by a combination of the smallest particles (amions); heat is the aggregate kinetic energy of the pre-atomic matter.

2. In my opinion, the generally accepted position that the planet went from the hot state of the plasma to the modern cool state through simple cooling in the cold space is erroneous. The gradual weakening of solar activity over billions of years is often called the cause of climate mitigation. Recognition of the thesis that climate change and climate formation as a whole is the result of external influences from outer space is a fundamental error at the level of fundamental physics. Physicists, and then climatologists, do not take into account that the solar system, including our planet, and nature as a whole are objects and simultaneously subjects of evolution.

The so-called **system theory of climate**, which includes a number of new ideas, was designed to prove the insignificance of the anthropogenic factor in climate processes. For example, its supporters try to calculate the amount of solar energy received by the planet per year. They introduced the concept of albedo – the ability of surfaces or individual bodies to reflect solar radiation – into climatology. It is believed that snow has the largest albedo – about 70-90% (the fraction of reflected radiation in the total volume of incoming radiation), which greatly delays its melting, especially in the Arctic.

When using the concept of albedo, one must keep in mind that after irradiation, substances produce a reactive return of heat to the atmosphere. The climatic effect is practically independent of color; it depends entirely on the density of the substance on which the sunlight falls. One well-known environmentalist was so keen on the albedo he suggested that all the mountains in the world be repainted in white.

The essence of this phenomenon is the reactive heat release from the denuded land: its amount becomes larger than the amount that came from solar energy. It's easy to make sure that the air temperature, say, on a platform covered with paving stones and on a nearby open lawn will be different; for example, on a sunny day, the air temperature on the platform will be higher than that of the lawn.

In fact, this phenomenon is much more complicated. It should be considered at the level of the atomic structure, which we will do later. Globally and fundamentally, the question should be put differently: whether our planet can accept and absorb solar radiation (kinetic energy)?

In my childhood, I was taught fighting skills, the aim of which was to use the force of the attacker against themselves. For example, if you are being pushed, then you need to pull the attacker in the same direction, and they will lose their balance and fall. In martial arts, historically, there is a whole system of wrestling that uses this principle, both in the literal and in the general philosophical sense. In the twentieth century, such a well-known martial art as aikido was created on its basis. Aikido techniques are based on merging with the enemy's attack and redirecting their energy. One of the versions for translating the hieroglyphs in this name is "the path of merging with life energy."

Nature acts on the same principle: it uses energy coming from the Sun for evolutionary development. As a result, and this is a paradox, nature uses living gentleness, intelligent power, and subtle natural technologies against the fire of the Sun. No matter how much we try, we will not be able to come up with a better version of sulde, or a way of sun protection, than the sulde itself. In the non-organic world, be it stones, metals, artificial materials, etc., nothing can withstand the long-term effects of solar radiation without detriment to its properties. Over millions of years, the wildlife has only become better.

When creating climate theories, scientists often begin to balance the heat and cold without taking into account the main thing – the reactive heat release from land. First, they calculate the amount of solar energy entering the planet's surface, then subtract the sum of the reflected rays, take into account the cooling of the planet by the space, and get the remainder in specific heat units. In other words, they calculate both debit and credit, deriving the difference of thermal energy, which is brought by solar radiation into the heat balance of the planet. Then this residual quantity is compared with the energy produced by humanity. We get a ratio of approximately 100 to 3; that is, it turns out that the energy produced by us is only 3% of the total debit.

Physicists have turned into accountants without noticing that nature is intelligent and perfectly organized. Over billions of years of evolution, it has managed to create an integral evolutionary system with sophisticated combinations of different levels of complex material structures. It has learned to 'cut off' the flow of heat to the planet and interrupt the 'electric circuit' with the help of living insulation, i.e. green cover, or sulde. In their physical essence, the sun's rays are electrical flows in the ether. Electric current in dense structures, such as metallic conductors, and electromagnetic waves of solar rays in ether have the same spiral-wave nature.

Let us dwell on the nature of the solar rays, which we often call solar radiation. As we have figured out in the previous chapter, they are formed during implosion processes in the solar plasma. They are literally discharged from it in the form of nanovortex impulses and reach a speed of light of 300,000 km per second.

It will be appropriate to supplement the well-known comparison of light with stretched springs. The pieces of springs are an analog of quantum in physics, or impulse. Physicists defined these concepts long ago, but they have yet to reach a common opinion on why light spreads impulsively. I see the reason for the quantization of solar energy in the fact that a quantum is a spiral nanovortex comprising of amions. Discharged from plasma, the nanovortex expands as a spiral spring into the amionosphere.

The pieces of springs are a chain of amions (photons) connected by aser. They are formed with the straightening of nanovortexes during implosion processes. Imagine that these springs are brought to Earth in the form of waves at the speed of light! At the same time, the sun's rays carry a maximum of kinetic energy and a minimum of pre-atomic matter; in space, electromagnetic waves (solar rays) propagate in the form of impulse waves in the ether.

Nature was able to put up the power of planetary consciousness and high organization against the effect of reactive heat exchange of land, or high non-efficiency factor, as I call it ironically. The non-efficiency factor turns out to be above one; that is, more than 100%, pretty much always, which is especially unpleasant. In the industrial era, the non-efficiency factor phenomenon returned from the distant past, manifesting itself in the form of catastrophic consequences of hurricanes and tornadoes.

We have already spoken about the primary evolutionary medium; the world was born out of chaos in it. The sixth stage of the physical evolution of the world has provided us with a primary electrotechnical medium in the proton of matter, where we can see both the efficiency and non-efficiency factor. Later, a primary bioenergetic medium was born, which began to conduct biocurrents – the basis of our biological consciousness. I'm sort of getting ahead of myself; we will talk about other types of medium later.

The middle part of the planet receives excess heat from the Sun, and the star itself is a perpetuum mobile that cannot slow down its speed; at least in the time frame of humanity. A low level of solar irradiation of the planet is difficult to imagine even theoretically. At the same time, there would be a real danger if the water cycle in nature is weakened. The level of solar radiation should be considered, taking into account the planet's flattened shape; regions of the Earth close to the poles would not be able to get the right amount of radiation.

The nominal level of solar irradiation can be considered 'divinely created,' or a jackpot winning in the global cosmological lottery (random system). There is simply no other option for existence. At the same time, our planet is protected from excessive harmful exposure only through the complex structure of sulde and its development.

The Formation of Near Space

The Earth was born as a huge gas protoplanet without satellites. There was a plasma core in its center, and it consisted mainly of primary matter with a small inclusion of light gases. The density of the protoplanet decreased in a geometric progression from the annular center to the periphery. It did not have a formed solid crust.

Over time, the gas planet began to take the form of modern Saturn: the central part got separated with the formation of a disk in the ecliptic plane. Inside the disk, a lot of small and large vortexes appeared – potential satellites of the Earth – which gradually merged into one large vortex, or the prototype of the Moon. The evolutionary meaning of the formation of planets and their satellites is the formation of entropic contrast in near-planetary space. As we have previously determined, the satellites of the planets are the third level of the vortex. Tertiary vortexes began to collect a loose part of the gas planet (a cloud of primary matter) into the body of satellites.

The central part of the protoplanet was compacted simultaneously with the formation of satellites under the influence of internal forces of attraction (aser). As a result of its transformation, two subjects of evolution – the Earth (the central plasma part) and the Moon (the former loose part) – appeared. As already mentioned, at the present stage of evolution, the Earth is a proton, the Moon is an electron, and together the planet and its satellite form a single atom. The Earth got an atmosphere and a near cold space; they are the prototype of a planetary atom. The entropy (amionic matter density) of the planet's crust and the entropy of the atmosphere (amionosphere) have a huge contrast; they differ by 1500 times or more.

<u>Note.</u> I call the prototype of the Earth the gas planet to suit modern cosmologists. In reality, modern gas planets mainly consist of pre-atomic (amionic) matter.

Correlating the modern outlines of Saturn and the remoteness of the Moon, we can imagine the dimensions of the gas planet at the time and the colossal amount of the evolutionary 'work' done to clean up and deplete near-Earth space, referring to the gathering of 'nutrients' by the loose part of the protoplanet for its now solid body. What we consider today as a near cosmos is the prototype of the inner atom once filled with rings.

I called this period of physical evolution **the formation of near space and atmosphere (amionosphere)**. In the future, evolutionary processes will occur mainly on the surface of the planet, on the border of contrasts. As a result of the amionosphere evolution, the **rhythmic variability of medium** appeared on the planet surface after the formation of near space. It occurred in the lower layers of the amionosphere; this understanding is very important because it was there that the drama of further evolution took place.

It was the surface of the Earth that became an isolated object and, at the same time, a subject of further evolution. The Sun got 'lit' in the planetary system somewhat earlier, and now the side of the rotating planet turned to it began to be affected by solar radiation – such is the condition necessary for the occurrence of rhythmic variability of the medium.

Deviating to modern philosophy, object-subject dualism is a fairly well-studied concept. In the following discussion, we will only disclose its evolutionary essence and dual nature. From the standpoint of evolutionary theory, evolutionary medium and its variability appeared on the surface of the Earth with the formation of near space. The variability of the medium arose under the influence of solar radiation superimposed on the rotation of the planet around its axis. The solar system should be considered the basic evolutionary system, the main subject (focus) of which is the planet Earth.

Evolution is a directed change of any process, object, or system toward complication, which has an irreversible character. In our case, the object of evolution is the surface of the Earth. This change always occurs in real, dynamic, or historical time. As a rule, evolution takes place gradually by accumulating micro-changes in the structure of the object, which, incidentally, play an important role not only in the physical and biological world but also in the public sphere – in particular, in science.

The surface of the Earth as an object of evolution acquires a new structure; land and world's oceans appear. As a rule, any structure has a double binary essence, an indissoluble internal contradiction. But together, the land and the world's oceans constitute the basic evolutionary medium; their joint quality is the basic climate determined by their great relationship. It should be especially emphasized that the climate is the quality of the evolutionary medium as one of the components of the evolutionary system.

The evolutionary medium is the primary concept; it is the accumulation of amions in the galactic space, the primary 'broth' in the water areas (life was born there) and the conditions of primitive people in the rainforest (that is where humans appeared). Later, each medium in the general hierarchy becomes an independent evolutionary subsystem with horizontal and vertical connections.

The development of the evolutionary amionic medium led to the formation of the Solar System and the planet Earth, where the basic climate appeared. Due to the development of the 'broth' of complex molecules in the coastal waters on the land surface, living nature arose. Then, on the basis of the basic climate, a new quality of the whole planet was created – a gentle climate and a biological life. In the paradise-like conditions, a person emerged from the animal world. Now it is consciousness and self-awareness that are the objects of evolution. The development of a new object takes place in a new evolutionary medium, namely the social medium.

The evolutionary system created the rhythmic variability of the environment on the Earth's surface, initiating the rise of the evolutionary movement to a higher biological stage. Rhythmic variability is the action of the entire Solar System and the rotation of the planets around its axis. The planetary system created daily variability; later, as the evolution of biological evolution evolved, epochal and annual variability of the environment appeared.

As a result, the vector of evolution aimed at creating favorable conditions for the evolutionary medium for the continuous complication of our world became clearly defined.

The Formation of Crust. The Formation of Atoms and Matter

The surface of the planet was a hot plasma – an extremely densified amionic matter. Under the influence of diurnal variability of the medium, vortexes of the fourth level appeared in the plasma. Then the microvortexes were transformed into chemical elements (atoms) and substances. The diurnal variability of the medium created the conditions for the next stage of evolution, namely, the creation of a homogeneous series of atoms and more complex structures.

The chaotic energy of the primal matter (the energy of hot plasma) is neutralized by the construction of atoms. This occurs when there is a combination of two a priori properties of the primal matter – inertia and aser – which can create vortexes. The natural vortexes arise in two opposite extremes, in two extremes of the physical medium: in a situation of weightlessness, or in a medium with extremely low entropy; and vice versa, in the situation of the highest density of primal matter. The second extreme is the plasma. I emphasize that vortexes are formed at the earliest opportunity under appropriate conditions. I imagine plasma in the form of lightning (electric arc). At the time, our planet was a huge clot of lightning.

The initial stage of planet structuring is the birth of light atoms and the emergence of an atmosphere. First, a boiling plasma emits microscopic vortexes into the near space (amionosphere), where they are immediately cooled, resulting in the production of light atoms (gases). The protoplanet has a cloud of amions; it is structured into a gravitational network. As the distance from the planet surface (or from the plasma) increases, the density of the amionosphere decreases in a quadratic function and becomes the cause of the decrease in the forces of gravity.

As part of the amionosphere, all the amions are engaged in a 'gravitational case'; they are 'regular workers' of the gravitational system. The initial density of the amionosphere, or density in the lower layers of the atmosphere, directly depends on the level of gravitational forces softening in the planet's crust. The thicker the crust, the poorer the amionosphere. But this advantage has its disadvantage: when solar 'electricity' hits the crust surface, dense atoms react with a sharp release of free amions.

I described the nature of the crust and the mechanism of its formation in the chapter, "The Formation of the Solar System." Outside the crust, the density of the amionosphere as the carrier of gravitational forces sharply decreases. The depletion of the amionosphere occurred during the evolution of planets and satellites; the satellites 'cleaned up' and created near cold space.

After the inclusion of light atoms in the composition of the amionosphere (clouds), the gas molecules begin to displace part of the 'regular' amions. The amionosphere becomes an atmosphere. The Earth's gravitational forces can retain only a limited mass of amions. So after the inclusion of the gas molecules into the amionosphere, a part of the 'extra' amions precipitates into the plasma. All these processes contribute to a decrease in the density of the amionospheric part of the atmosphere, which immediately leads to a decrease in its temperature.

The high density and, as a consequence, the temperature on the planet surface can be compared with the chaos of things scattered around the house. If you sort it all out in cabinets and suitcases, you will introduce order in your home and get an extended habitat – a 'cooled' living space. The things scattered around the house narrow the living space: the constant or temporary chaos of amions and the excess of primary matter in the atmosphere (amionosphere) directly increase the temperature in its lower layers. As a result of the appearance of light gases, amions end up sorted out in the 'cabinets' of molecules and atoms. More precisely, the molecules of gases replace the free amions in the lower layers of the atmosphere. The climate of the planet begins to form.

I explain the formation of light atoms not by nucleosynthesis (a thermonuclear reaction), but by a process that I call "the effect of a coal emitted from a fire": the vortex plasma clot of amions ejected from the disk of the sun (or plasma of the planet) compacts, or 'cools down' in the relatively cold part of the amionosphere, and forms light atoms.

In modern physics, the emission of radiation energy from the interior of the Sun is considered the result of a thermonuclear reaction – the synthesis of light atoms into heavier ones and the formation of atoms in the interior of the star. I proceed from the antithesis: on the contrary, it is the formation of atoms that 'cools down' the medium.

Let's go back to the plasma state of the planet. After the formation of the atmosphere (displacement of 'regular' amions from the amionosphere), an imbalance of low entropy of the atmosphere against the highest plasma density was formed. Because of the abrupt transition in density levels, the plasma tried to form a film in the boundary medium. This process is similar to the formation of foam on the surface of boiling milk. The plasma body of the planet was fully ready to create a film on its surface.

All of the above can be rephrased as follows: the surface of the planet was ready to stabilize and create a materialized border with a depleted medium. Crust formation is one of the most important processes of evolution.

Secondary vortexes form inside a large space vortex, a former ball. Now we can distinguish two types (level) of vortexes: a large vortex with a central star and internal vortexes of the planets. Then planets and satellites collect (clean up) amions in their orbits, and then a depleted interplanetary space – a cold cosmos – is born. We can say that subordinate small systems are born within a large system, and that a sharp contrast in the entropy of the medium forms between them. In turn, the sharp contrast contributes to the formation of the crust (film).

The above abstract ideas can be confirmed and consolidated with simple and understandable examples from the surrounding world or from history. In this case, firstly, we will better understand physics and nature, and secondly, we will be able to know history better. Therefore, the figure of Genghis Khan will now appear in the discourse.

In pursuing his grand conquest, Genghis Khan defended the looting of captured cities traditional for that time; lasting up to three days, the looting was the release of malice accumulated in the soldiers or, in our terminology, the release of kinetic energy. At the same time, Genghis Khan was always interested in the culture of the conquered countries; he tried to unravel the meaning of its various manifestations and apply the finds to good use.

Once upon a time, during a military campaign in China, the advisers recommended him to contact a famous Chinese philosopher and a carrier of advanced ideas. When proposed to join the conqueror's service during their meeting, the philosopher demanded the immediate cessation of violence and looting in his city. Genghis Khan accepted the condition and gave the appropriate order, doing it contrary to the tradition established in the world, including in his army.

We should pay special attention to the speed of execution of his order. The higher commanders (*temniki*) gave commands to their ten thousandths *tumens*, junior commanders gave orders to thousands, *sotniks* to hundreds and, finally, *desyatniks* gave orders to every soldier. As a result, thanks to the fact that the troops had a hierarchical and manageable structure, robberies and violence in the city, for which the philosopher interceded, quickly ceased.

This story shows us not only the effectiveness of the organized structure but also the process of taming the wild malice and the kinetic energy of the warriors, transforming the chaotic energy of the unorganized mass into multistage vortexes and their complex hierarchy, which was the brainchild of Genghis Khan himself.

He created a hierarchical, easily manageable military structure, putting people who showed special knowledge of military affairs at the head of the largest units. Before the army reforms of Genghis Khan, Mongolian troops consisted of different tribal groups of different numbers, which were ruled by tribal leaders who did not always possess the talents necessary in military affairs, and the tribal union itself was based on voluntary principles. This often led to unpleasant surprises; for example, some leaders could lead troops away before a decisive battle.

But this is already in the past; now we can see how the hierarchical order of complex structures can curb the wild energy of chaos. Summarizing, we can say that this pattern extends to the whole world from the motion of atoms to the processes taking place in human society. Self-structuring and complication of matter into atoms, molecules, and substances led to the first taming of hot plasma and its transformation into cold atoms.

As a result of the natural structuring of the planet's crust, two large and sharply contrasting structures emerged: the land and the world's oceans. Climate mitigation and stabilization of the atmospheric phenomena have occurred as a result of the physical evolution of the world. I also call this the **hot evolution** – the geological epochs of atom formation. The formation of the lithosphere, and then the world's oceans, began to gradually soften the already formed integral climate of the planet. The powerful geological bursts began to gradually wane and weaken.

There was a great dialectical balance of land and the world's oceans. Biological forms took up the formation of the climate, gradually covering the surface of the land first with primitive species, and then with complex ones, including forests.

Miracles were happening on the planet: the climate became a completely controllable evolutionary process! The new Evolutionary Theory of Climate leads to an understanding and confidence that we will be able to manage the climate almost manually, if we choose the right way to develop sulde and do our best to protect nature.

The planet's crust, or a compacted boundary of a complex structure with strong entropic contrast, was not created in one moment; this happened gradually as a result of the solar system structuring as a large vortex in the form of a uniformly distributed medium of amions. A compact center appeared in the particle cluster, which then compacted the medium. All these processes can be called the structuring of the chaos of the medium into certain forms of motion. Basically, any material structure is a form of particle motion, a dynamic system, and a dynamic system is the equilibrium of forces of the subjects of the medium.

In terms of physics, plasma medium is the chaotic movement of unorganized particles (amions) in a confined space, or figuratively speaking, in cramped conditions. The boiling plasma medium consists of large Coriolis vortexes, which, in turn, consist of microvortexes of future atoms. When this medium is heated during the daytime, there is a sharp increase of them in the plasma of the planet (day), and during the cooling (night), the distance between them increases. Then the microvortexes transform into atoms. Upon cooling, the contents of the vortex plasma compact; aser does not allow them to expand, as a result of which the atom gets structured.

The relevant question then is: what qualities of the atomic structure create its stability and where did the fundamental forces of interaction in the nucleus of the atom come from?

Firstly, amions, with their properties of inertial motion and attraction (aser), cannot be organized into a complex structure except in a vortex way. Why does the atomic vortex become extremely stable and does not dissipate, like the steppe one? The point is that with a sharp cooling of the vortex, the atom is structured, increasing in size. The atomic vortex becomes multilayered; each layer has the same amount of amions, and each amion has three nuclear bonds: two in its own layer, and one more in the outer layer of amions. There will be much more amions in the last layer of the atom, and it becomes the atom's crust; the atom itself becomes a systemic construction of amions in which they are all busy with work, and their talents are in demand. They are in a continuous and perpetual circular motion, with all of their aser properties being used. See Fig. 29.

The atom is like a rotating carousel, the construction of which is similar to a three-dimensional farm, where the nodes are the amions bound by aser. When the carousel rotates, they realize their motor activity, and the aseric properties ensure the stability of the carousel (farm). I emphasize that this is only the initial stage and the general scheme of stability; in fact, the atom and its construction are much more complicated.

Secondly, we said that within the atom there is a redistribution of forces. The amions of the atom's outer layers begin to move faster. They receive part of the kinetics from the internal (nuclear) part and become negatively charged. Accordingly, the amions in the nuclear part have more aseric properties than kinetics, and they receive a positive charge. See Fig. 13. At the center of the atom, the amions have less motor properties than aseric ones; the conditions for Avogadro's law change: now more amions can fit in a unit of volume. This phenomenon and layer-by-layer structuring of the atom make it an extremely stable structure.

We have talked about the atom as a spatial farm, and this construction design allows to cover large areas. It has nodes fastened by ties. The farm becomes super stable if it is shaped like an arch. As a result, the consumption of materials in the construction industry sharply reduces. Our atom has a round or an ellipsoid shape. Taking into account the nanoscopic nature of the amion and the colossal dimensions of the atom in relation to the amion, you can imagine just how cunning nature is, and what kind of material it uses for the atom stability. A material that can split and stick together; this is what makes it different from a rigid construction farm. At the same time, the atom is a dynamic structure; it is a self-moving carousel and also a natural automaton with a primitive consciousness. In complex structures, a set of primitive consciousnesses constitute an algorithm of complex consciousness. The atom is a cell of inanimate nature; it is also the basis for the development of living nature.

In the local plasma (evolutionary) medium, irreversible processes occur: first, with an increase in the kinetic energy of the plasma, the chaotic motion of the amions is transformed into a multitude of nanovortexes. Then, when cooled, the nanovortexes form atoms.

It should be noted that the plasma (evolutionary) medium could not lead to the formation of atoms if it was not for the influence of the system factor. The evolutionary medium began to 'bear fruit,' thanks to the Solar (evolutionary) system, which creates the variability of the medium – the situation of heating and cooling.

That is where the concept of the irreversibility of evolution is born in the physical sense. The Solar System unclenches the atom, and then compresses it; aser prevents it from unclenching again. Let us recall the famous saying of Heraclitus that one can never step into the same river twice: the next day, the amount of atoms increases, and after a while, a small crust of atoms appears on the plasma surface. And this is a different plasma river altogether!

When the medium is sharply cooled, the nanovortexes of plasma matter do not return to the form of a simple plasma; the set of nanovortexes does not dissipate, but immediately becomes a set of atoms, that is, matter. Atoms are not formed separately, but in the form of an alliance, since the aser of microvortex centers stabilizes them in a specific form and they are born on a massive scale in the form of substances, or a union of atoms. There is a layered formation of the planet crust.

The formation of the crust is a trend. If there is a contrast of entropy in the structural system, then a crust, or a formalized boundary separating different density levels must necessarily appear on the boundary between the plasma of the planet and the amionosphere.

What would happen to the plasma if it was not for the variability of the medium, which is created by the rotation of the planet? With continuous heating, the plasma would boil like melted steel in a blast furnace, which continuously operates without unloading. To obtain blanks of a given shape, cooling is required; in other words, the variability of the medium will also be needed.

The transformation of the melted plasma of a protoplanet into atoms and substances occurs during the cooling of the medium. In the future, the very fact of the formation of crust from material structures leads to a decrease in temperature on the surface, while the amplitude of temperature oscillations under the planet's crust increases. It can now be compared with a boiler covered with a lid. But this cover will have drain valves – volcanoes and geysers – and this is already our time.

Implosion: The Stick Effect and Quantum Mechanics

In the chapter, "The Formation of the Solar System," we considered three types of vortexes (three levels of vortexes): a large vortex, a planetary vortex, and a vortex of the primary matter in the formation of planet satellites. Now, I want to explain how internal vortexes form using a good example. In the text and in the drawings, I will call this phenomenon (process) **the stick effect**.

People who use a whip (stick) know that if one cannot use it the right way, they can, with a sweep, knock out their eye with the end of the whip. To make a strong and accurate blow, the person hitting with the stick must be able to swing right – i.e. back and round. At swing, the end of the whip (stick) is twisted in the opposite direction. This is the effect of the Coriolis force (see Figures 21 and 22).

Figure 25 shows us the formation of internal vortexes of amions (fragments of frames). The processes of internal vortexes develop and move into a phase, which I called **implosion**. Implosion is an inwards explosion. We have already considered this phenomenon in connection with the sun radiation: under the influence of the Coriolis force at the level of micromechanics, multilevel vortexes are formed. In the solar plasma, the vortex processes can develop inward in the format of the fourth and fifth levels. It is very difficult to speak about the nature of implosion processes, but it is obvious that there are such phenomena in nature.

What happens in the bowels of the Sun? In its extremely dense plasma, multilevel vortex formation takes place up to the formation of nanovortexes.

Turbulence of the Coriolis force in plasma matter periodically ejects vortexes in the form of microscopic plasma clots into the surrounding space – into a medium with low entropy. After that, ultra-light atoms (hydrogen and helium) form around the solar disk, but they do not last long – the forces of attraction of a dense mass attract and absorb light atoms back into the Solar plasma.

The concept of implosion leads us directly to quantum mechanics, namely to its principle of uncertainty. At the present stage, we cannot, with the help of special tools, observe how implosive processes occur inside the atom or in the bowels of the Sun. The empirical method of cognition has reached its limit; we are left with only metaphysical methods and indirect facts (Freud's method). Trying to explore the microcosm and penetrating it with the help of various rays, we thereby change the already structured medium.

The principles of implosion processes occurring in the microworld are common to different sciences. The details and nature of these processes can be described as follows: we know that we do not know anything and, probably, we will never know. At the same time, we know with maximum probability that the atom is stable, because we know the principles of its construction: the vortex character and the double nature of the amions.

This can be compared with the work of meteorologists. Neither 100 years ago nor now are they able to absolutely predict what the weather will be like, because the detailing of its components occurs in the microcosm of atoms – the amionosphere: cyclones are big vortexes, and anticyclones are the structuring of its tropospheric part. Fixing temperature records from time to time, meteorologists know for sure that the average annual temperature of the planet will change insignificantly.

Variability of the Medium is the Cradle of Evolution

The evolutionary medium transformed into a complex structure of the Solar System in the form of a star in the center and planets with their satellites on the periphery. Born as a result of evolution, they become independent subjects of evolutionary processes.

Planets and their satellites experience a constant change of warm-up and cooling, which means that a rhythmic diurnal climate appears on their surface, leading to the formation of atoms and substances on their surface. Medium and heavy atoms can be born only in a complex way: many atoms, at once, emerge from the hot plasma. Then they are structured as substances. The path, predetermined by universal evolution, is the continuous complication of matter.

The combination of the inertial motor energy of amions (active principle) and their attraction forces (organizational beginning) lead to self-complication of matter. We have already said that the Solar System is a large and integral evolutionary system, and the large system creates a situation of medium variability in its internal subsystems. At the final stage of the physical evolution of the world, the entire Solar System becomes a huge 'factory' for the production of atoms on the surface of our planet and other cosmic entities.

If at zero time, the transformation of the primary matter in the Big Atom occurred arbitrarily, then a 'space plant' is created on the surface of planets for the formation of atoms and substances. Practically autonomous processes of self-complication of subjects (amions) in the primary evolutionary medium now become a single large-scale process of the evolutionary system, taking place in a complex and accelerated, compulsory manner. The formation of a homogeneous series of atoms and their combinations in the form of thousands of species is the hot stage of evolution, or simply the hot evolution mentioned above.

Since that time, the formation of the planet's solid crust begins; there is an increased drop in entropy (and temperature) between the protoplanet crust and the amionosphere (atmosphere). Earlier, the density of the planet smoothly decreased from the geometric center to the periphery, although this smooth transition occurred in a quadratic function.

Heavy and medium atoms are born immediately in the form of matter; each atom is not born separately. A bond is formed from free amions in the interatomic space between microvortexes (atoms) – the union of atoms becomes a substance.

To understand what a substance is, we will consider its antithesis – a solution medium, in which particles can be in a suspended state. Each atom or molecule in the solution has its own crust. They do not have their own crust as part of substances; they share a collective crust. For example, if we consider water as a substance, then it will have a common crust for all molecules – a water film. Individual molecules will not have a crust. But molecules of salts dissolved in water have a separate crust, which is why they easily precipitate. Only with the evaporation of water, each of its molecules forms its own bark.

If the atoms of gold are represented in the form of balls, the space between them is filled with free amions (gluons). The adjacent atoms have common binding amions; at the same time, it is as if they glue them together with chemical (aseric) bonds, forming a substance. It consists of atoms and their interatomic bonds, which together constitute the common protobody of matter. The bond between atoms is realized by forces of mutual attraction through amions in an interatomic medium.

The chaotic motion of particles in the plasma on the surface of a protoplanet is transformed by the Coriolis force into a vortex motion. This process increases sharply upon irradiation of the planet's surface; implosion processes are accelerated and reach the stage of nanovortex formation. Then, when the medium is cooled, vortexes are stabilized in the form of atoms, which in this context is the result of metamorphosis of plasma or substances. We have already decided above on what is matter.

The process of the creation of atoms can be visualized by observing how the liquid clay is dried up. With accelerated drying, it breaks into small fragments. First, the clay is diluted and 'dilated' with water. When heated and evaporated, it contracts, and the clay grains converge with each other by gravity. Then it is divided into fragments. We can say that fragmentation occurs due to a sharp change in environmental conditions. We note that the metamorphoses of clay are reversible after a new moistening. The atoms do not break up into amions with a new heating, as the process of formation of atoms is irreversible.

The processes of 'self-assembly' of atoms are the processes of structuring a set of amions in the form of complex dynamic formations. Each atom becomes an autonomous and stable vortex.

After the formation of the first crust on the plasma surface, the temperature in the amionosphere immediately decreases and the medium is cooled because of the 'packing' of the plasma energy in the atoms. A non-discrete plasma medium is an extremely dense pre-atomic matter. In any part of it, the particle density will be uniform, and the temperature will be above 2000 degrees Celsius; the plasma will be very hot. The density inside the atom is uneven. In the structure of the atom, the predominant part of matter is in the nucleus, and the dimensions of the nucleus are extremely small in comparison with the atom itself.

The energy of the amions (the plasma composition of the nucleus) is hidden deep inside the atom, so we can touch the 'tamed' plasma in the form of substances. Atoms are like candy in a multilayer wrapper. The wrapper itself consists of a multilayer primary matter with a density decreasing at the periphery.

At the beginning of the path, the engine of the hot physical evolution is the diurnal variability of the medium. The diurnal changes form a solid crust of the planet from refractory minerals very slowly, bit by bit. The protoplanet gradually becomes a planet, which has a very thin but already solid crust made of substances. The surface is no longer a fire-breathing plasma; the plasma energy on the surface of the planet is neutralized in the materials of the crust or, more accurately, inside the atoms.

The plasma is covered with crust or even a flexible film of atoms. It has become protected, and now the solar radiation cannot immediately initiate implosion processes. It takes time for the planet and its crust to warm up. The thicker the crust becomes, the more time it takes. The physical stage of evolution begins to take place within the framework of geological epochs.

At first, the epochs were very short; later, they extended to millions of years. The extension of epochs is associated with the thickening of the planet's crust. The daily cyclicity of warming remains, but at the same time, there is a geological cyclicity of the climate. The two-level variability is created on the planet: in days and epochs. After the formation of the solid crust, the planet first appears as a vessel, and later as a closed thermodynamic system.

When heated from the outside, internal stress accumulates in this thermodynamic system. At a critical moment, this voltage leads to plasma eruptions, and then, when the plasma is cooled, a new cycle of formation of substances takes place. Under the new conditions, micro- and nanovortexes arise already under the planet's crust, and their stabilization – the formation of substances – occurs when the plasma leaves the surface of the crust. At the same time, as a result of sharp cooling, new atoms and new substances are born again. Gradually, the planet's crust thickens, so more time is needed to pass the geological cycles.

Much later, after sufficient enhancement of the crust, geological eruptions begin to occur less frequently, but at the same time, they occur in an extremely strengthened form. At such a time, dust and gases rise to the sky due to a number of eruptions. They cover the whole sky, as a result of which the access of solar rays to the surface of the Earth is sharply limited. Because of the sharp cooling of the amionosphere, the planet's shell shrinks considerably and the so-called nuclear winter sets in. Introducing such a name, scientists of the second half of the twentieth century were able to primarily see the climate threat of the consequences of a large-scale nuclear war.

When I use the phrase "the planet is cooling," it does not need to be understood literally: when the solar radiation decreases, the planet enters into an entropic equilibrium only with the cold space. At this time, it begins to produce the strongest assimilation (absorption) of amions from the amionosphere, the medium (near space) becomes even poorer and colder. Cold brings cold.

We know that with rapid cooling, dense material structures shrink. Due to the cold in the amionosphere (atmosphere), the shell of the planet shrinks in volume and compresses the inner plasma content. At such times, the planet's crust may crack due to the high pressure of the internal plasma. The division of Pangea and the movement of the continents could subsequently take place precisely for this reason.

Gradually, the sky becomes clear and the planet heats up again; however, it takes too much time for a critical heating up of a planet with a fairly thick crust, which is called a **geological epoch**. As already mentioned, after the appearance of a thickened crust, the planet becomes a completely closed thermodynamic system. With a strong heating for a long time, a colossal energy accumulates inside it. When the critical state is reached, the contents of the planet explode again.

In place of the dominance of diurnal and multi-month variability of the medium, very long **epochal variability** comes in. This can be assessed as the emergence of a long 'peaceful' period of development, or figuratively speaking, the period of evolution between revolutions.

Such geological processes are not just cooling of the Earth in space. There is self-structuring – the complication of matter on the surface of another hot planet. Later, water is formed, which is squeezed out as a vapor from the planet's body, and then condensed on its surface mainly at the poles.

The equatorial part of the Earth's surface always heats up more strongly; therefore, its middle part undergoes frequent and strong thermodynamic transitions. Circumpolar regions receive solar irradiation at an angle; correspondingly, the level of incoming solar radiation per unit area of land in these areas is much lower than in the equator.

In the middle part of the planet, geological processes proceeded under the conditions of the strongest variability; therefore, it is there that we see all the major mountain formations. The Earth's structuring lasted for billions of years. Water constantly evaporated from its body – water surfaces appeared on the planet surface.

At some point, geological processes began to slow down; the ratio of land area to water areas was stabilized. The main reason for the slowing down of geological processes was the relative stabilization of the climate, which was the result of the planet structuring. In the process of physical evolution, the area of the water areas increased, which led to gradual mitigation and stabilization of the climate.

We have already called the dialectical and harmonic ratio of land area and water areas the structural basis of the first, or basic, level of climate.

Climate stabilization slows down geological processes. The planet, which was later called the Earth, fell into a special focus of evolution – the development of the biological world began. The formation of various biological forms, in particular microorganisms, occurred in parallel at the last stages of the hot evolution. The early development of microorganisms is evidenced by the fact that some modern viruses can perfectly exist at high temperatures and in a different aggressive medium. In the late stages of evolution in the world's oceans, a very complex biological life was actively developing in order to subsequently crawl up on the shore.

In the early eras under the conditions of high temperatures in the equatorial part of the planet, the normal water cycle in nature was objectively impossible. By **the water cycle**, I mean the new cyclicity in nature: the evaporation and condensation of water. We will talk about it in more detail in the relevant section of the chapter.

Climatic conditions that allow the water cycle in nature appeared only in the circumpolar regions due to the angle between the ecliptic plane and the axis of the planet's rotation. Hereinafter, I will call this angle **the tilt of rotation axis**. The planet's ecliptics is the plane in which the orbit of the Earth's rotation around the Sun is located. The given data of the planetary system have created objective conditions for a new round of biological life development – the release of biological forms on land in the circumpolar regions, about which we will also talk in more detail later.

Annual Climate Cyclicity

The annual climate cyclicity is associated with the tilt of axis of the planet's rotation round the Sun, which is 23.44°. Such an angle has proven to be the most optimal for biological evolution since it provides active irradiation of circumpolar regions and leads to a 'build-up' of the climate during the year – the Northern and Southern hemispheres of the Earth alternately pass the winter and summer period. There is a climate variability on the geographical latitudes of the planet and a periodic change in weather phenomena in an annual period in all climatic zones.

For example, the annual cyclical nature of the tropical climate is characterized by an annual onset of the rainy season in the tropics. Paraphrasing the figurative comparison from the previous section, we can say that the planetary system is rocking the biological life in the weather's annual cradle. The magical – and it is no exaggeration – tilt of the planet's rotation axis first awakened life in the circumpolar regions. This is exactly where the history of crawling up on land begins, which is confirmed by numerous studies.

Let us consider the origins of tilt of the planet's rotation axis. During the formation of water areas and land, its most part was closer to the North Pole, which led to a gradual shift of the planet's center of gravity. Rotating around the Sun, the Earth reached its maximum orbital velocity at the perihelion point.

Due to the action of the centrifugal force, the Earth's rotation axis shifts because its new center of gravity is already above the ecliptic plane. After the gradual displacement of the rotation axis, the center of gravity and the total vector of the planet's centrifugal force intersect at one point in the ecliptic plane. When studying Figures 32 and 33, we notice a simple and ingenious find of nature: the tilt of the axis expands planet zones. It stretches them without doing it in the literal sense – physically. Thanks to this, the areas suitable for living are expanding.

As we see, the evolution of nature expands the potential habitat for all living things, and the surface of the Earth gets the annual variability of the climate. The area of the 60th parallel of the northern latitude receives a sufficiently strong solar irradiation for six months due to the tilt of the planet's rotation axis. The same happens in the Southern Hemisphere.

Nature makes life possible in the circumpolar regions for 5-6 months. The necessary daily temperature difference is established, crossing the dew point – the temperature at which condensation of water vapor occurs. The tilt created a temporary strip of several months in a year, suitable for a 'comfortable' life between the cold poles and the heat of the equatorial part of the planet. When this area shifts into the cold, ceasing to receive the necessary amount of solar radiation, winter comes, which the biological world lives through in various ways.

Thus, the initial impulse to the emergence of life is created by a change in the temperature of the medium passing through the dew point, which sets the water substance in motion. In turn, the movement of water awakens and stimulates the metabolism (necessary for the maintenance of life in the body) of the proto-organics, or various microorganisms.

Biological Life's Exit to Land

In this book, I do not describe the origin of life; I will only note that it originated in the water medium. Now it is important for us to consider the influence of life on the climate. So first of all, we will pay attention to its exit from the water areas to land.

In the early geological epochs, the primitive life on land first appeared and developed in the circumpolar regions. The biological life's exit from the ocean on this relatively small area is due to the fact that the active formation of the planet's relief continued in the middle zone of the planet: it was hot, dry, and troubled. Evolution saved time by awakening life in the only possible place on the planet.

In the early epochs, the climate in the circumpolar regions was highly variable due to the small mass of the planet's biosphere and the weakness of its stabilization factors. Therefore, spontaneous fluctuation processes predominated in the near-earth atmosphere. Because of climate instability, biological species in the early eras could only be primitive and unproductive. They moved into the phase of waiting for favorable conditions for six months while wasting most of the vital energy to adapt to the harsh and unpredictable environmental conditions.

The dialectic meaning of the complication of matter and the development of life on land consists in mitigating the Earth's climate and mastering its hot and dry middle belts. The subjects of evolution adapt to the level of variability of the habitat and, at the same time, affect it simply by their existence. Biological evolution has always fought with the excessive heating of land; it gradually curbed and subdued the sharp changes in the weather regions, overcoming the dryness and heat of the planet's middle belt.

These processes occurred in epochal climatic cycles. The living nature adapted to the planet's conditions, protected the dry land from the direct rays of the sun with a green cover (sulde), and created better conditions for the water cycle in nature.

Basic thesis of the Evolutionary Theory of Climate No. 3. *The planet Earth initially gets an excessive amount of heat from the Sun. It is constantly under the threat of global warming.*

The possibility of global warming is affected by two objective factors of the planetary system: the uneven irradiation of the planet's surface and initially excessive brightness of the Sun, without which, on the other hand, we would not have a sufficient level of water cycle in nature. Thus, the excess capacity of a vehicle is in demand at a sharp acceleration or on a steep rise, but is not used in normal traffic conditions.

The Magic Properties of Water. The Water Cycle in Nature

The middle stripe of the planet was dominated by heat, so biological life originated in the circumpolar regions where the water cycle in nature was possible. Then the living organisms conquered the rest of the land little by little, for hundreds of millions of years. Drought-resistant plants – xerophytes and succulents – were in the forefront of biological evolution.

They are characterized by a quick adaptation to sudden changes in the amount of water in the environment. For example, succulents have special 'reservoirs' for water storage in their tissues. Some types of xerophytes survive extreme periods in the form of **seeds** and **spores** that grow after precipitation. Sometimes they manage to grow up, blossom, and give new seeds, which will long remain in a state of rest until the next rainy period, in four weeks.

Biological evolution on land began with launching grounds in the circumpolar regions, gradually covering the land with woodland and saturating it with water.

The natural watering and landscaping of the land is a binary opposition in the evolutionary movement. As the land was covered with forests and savannas in the course of evolution, the water cycle in nature was stabilized in parallel and in portions. Thus, the climate of the whole planet was mitigated, initiating the development of sulde. Biological evolution gradually suppressed the heat in the equatorial part, the average temperature of the atmosphere decreased, polar and mountain glaciers were formed, and great river systems and internal reservoirs were formed.

The formation of the water cycle in nature had been initially predominated by random factors; then, as the climate stabilized, atmospheric precipitation also gradually stabilized in different climatic zones.

The ability of water to decrease rather than increase its density when freezing is vital for the entire biosphere. When going into a solid state, the molecules of a liquid substance should be located closer, and their density should grow. This is how ordinary substances behave, but water is an exception. If water is gradually cooled, we will see an entirely normal process: it will become denser, and we will not notice any deviations from the norm.

However, when reaching the temperature of 4°C or below, water suddenly becomes lighter. Freezing, it expands by 9% in relation to the previous volume and forms ice, which does not sink, but floats. Such an expansion may be fatal for a water pipe in the event of unexpected frosts: freezing in pipes, water will rupture them. The water is characterized by viscosity and high surface tension.

If water behaved like an ordinary physical substance, rivers and reservoirs would freeze completely to the bottom in winter. Ice is lighter than water; it forms on the surface of water areas and becomes their thermal insulation. For example, at an air temperature of -20 °C, the temperature of water under ice does not drop below 0 °C in a frozen river. In winter, the life in the reservoirs goes on.

One more property of water should be noted: under the influence of sunlight, it evaporates easily from the surface of water areas; clouds form in the atmosphere. They partially delay the sun's rays, limit the access of solar radiation, and weaken the heating of the planet's surface. Water is a natural regulator, which is responsible for irradiating and heating the Earth.

For this reason, the islands and coasts of the continents are directly influenced by the climate of the oceans. The influence of the world's oceans on the climate of the central parts of the continents is limited; only greening and the presence of inland water bodies can influence it. The climates of such regions are called continental or sharply continental. In summer, it is hot and dry, and in winter, it is very cold.

Water is a magical substance, but life itself is no less magical. Together, they transformed the planet. This tandem has already been called a binary opposition. But the original driving force of the planetary system is a rhythmic effect of heat and cold on the planet's surface, resulting in setting in motion of all natural processes on Earth. Solar irradiation evaporates water from its surface, the ascending streams of air raise water vapor into the atmosphere, the space cold cools the water vapor, and atmospheric precipitation occurs. Gravity organizes the movement of water along the terrain, providing the water cycle in nature and, as a consequence, life on the planet Earth.

Coincidences in Nature is the Law of Evolution. Climate Stabilization

At the initial stage of biological evolution, atmospheric phenomena had a random character; they brought only wandering precipitation, which could fall out in any area of the land. The surface of the Earth was mostly stony. It took billions of years for water, wind, cold, and heat to destroy the stone, preparing 'granular soil' for future biological evolution.

Absorbing into the pores of hard rocks of the continents' rocky structure, water gradually destroys them. This is one of the manifestations of water's magical properties. For example, we see how the dense structures of stones and concrete products crumble during the freezing process. Wind and flow of water carry the formed elements and various minerals along the mainland, forming the soil. Its cultural layer begins to form in the lowlands; on the plateau, it will always be thin.

With daily freezing and thawing, water loosens the soil, including stony soil. People notice this kind of natural phenomenon in spring. Loose soil is better at retaining moisture. Every day, the Earth absorbs thawed water like a vacuum pump, swells, and prepares for the birth of a new life. It is as if nature conducted agrotechnical measures, automatically performing weeding and watering. The roots of plants become a frame for loose structures and prevent wind erosion; various organic deposits fertilize and enrich the soil. In improved prepared conditions, more productive biological species appear.

These processes took place over hundreds of millions or billions of years. It should be kept in mind that all evolutionary processes occur in parallel and simultaneously; there cannot be any preparatory period. The gradual change of some structures affects the development and generation of others. This is the main way of the evolution of nature in general and the climate in particular.

Strong thermal effects from space bring casual atmospheric precipitation to any regions of the planet. The plants resistant to dryness (xerophytes and succulents) come next; they fix and shade the land, protecting it from the burning rays of the sun. Then water randomly comes in, and as a result, biological life naturally stays in these areas.

The constantly recurring set of random phenomena is the tendency of the globalization of randomness, the probabilistic law of nature. Darwin's theory is based on this understanding of randomness. In general, the evolution of nature develops through trial and error. At the same time, nature involves all possible options, the best of which take root and develop.

State of chaos is a big spoon of evolution that stirs the climate kettle. It boils, and all its contents are 'languishing' inside. Since we are inside this boiler, we often think that weather and climate are random processes. In reality, nothing happens by chance; the randomness of weather manifestations is its regularity.

The climate over the year is random; it is composed of a perfectly tuned mechanism. The average annual temperature of the Earth was unchanged to a thousandths of a degree. This phenomenon led scientists to understand that the planet is a single, integral living organism organized from within. Nature regulates its temperature like a warm-blooded animal.

As the climate and climatic zones form, random atmospheric phenomena occur less and less frequently, and this is a new quality of the climate. It is associated with a decrease in the average temperature of the planet (we will elaborate on its role in the ETC below) and the frequency of strong warming up of continents and water areas, which leads to a weakening of cataclysms in nature.

As a result of a decrease in random climatic processes, stable climatic zones, differing in their remoteness from the equator and the oceans, the degree of greening and water saturation, as well as orography, appear. Climatologists use the term orography for the dependence of climate on the terrain. For example, alpine meadows appeared on the leeward side of the mountains, and they are oases inside the semi-desert and desert regions in the basins.

Conditions improve in new stable climatic zones, more complex biological species appear and develop, the amount of water increases. The regional climate is getting milder and stable regional water cycles in nature are formed. Parallel and continuous development of climatic processes and the development of biological species, their portioned continuous 'mutual aid' is the dialectic of the development of nature; it is evolution itself.

The evolution of inanimate nature creates only basic climatic conditions and the primordial nature of biological life. It is the biological evolution, coupled with the continuing physical evolution of the world, that leads the climate to a gentle level, which we will consider further. It's worth repeating that this is why the proposed theory of climate is called evolutionary.

Biocenosis. Acceleration

Biological species are in interspecific cooperation, which is called biocenosis. Biocoenosis is the achievement of biological and general evolution, complex hierarchy of biological diversity in nature, coexistence of plant and animal worlds – flora and fauna – as well as the intermediate forms of life, such as unicellular and microorganisms, fungi, etc., in a single living space.

I want to emphasize that biocenosis in living nature is a large and integral biological organism that is formed in a certain area of the terrain under favorable environmental conditions. In the evolutionary theory of climate, we go from the concept of local biocenosis to an understanding of the planetary unity of living nature and the physical structure of the Earth.

Biocoenosis is organically inseparable; it is constantly being improved and complicated, as a result of which the environmental conditions rapidly and intensively improve. Its creation and maintenance, the evolution of biological species, are inconceivable without biocenosis. The planetary biocenosis is a 'biochemical combination' of living nature on the planet's scale; at the same time, it must be taken into account that it depends entirely on the inanimate nature and the climate. During climatic and other crises, local biocenoses die completely because of their systemic complexity.

One of the examples of the highest form of biocenosis is the tropical forests, in which ideal conditions for living nature are created. A significant part of the biological diversity on Earth exists precisely in the tropical forests as the counteraction of a large number of cooperated wildlife to strong solar irradiation. It fights against it using its organic nature – a stable internal connection of subjects in the evolutionary medium.

But the best conditions are created by nature itself. The essence and content of a favorable climate is life itself, as well as the existence and self-development of the biosphere. For example, if we destroy the rainforest, we will immediately change the climate, because under the influence of strong radiation, this area will turn into a desert or semi-desert. The tropical forest as the highest form of biocenosis creates the prerequisites for anthropogenesis. The tropical forest gave security, shelter, microclimate, a highly productive food base, and 7 million years of successful development and acceleration to the future man.

The phenomenon of **acceleration** shows that the born individuals living in comfortable – compared with previous generations – conditions spend less life energy on adapting to the medium. It is aimed not so much at survival, but at the development of the body, its morphological features. I attach the expanded meaning in the context of general evolutionary logic to the term acceleration and extend it to all self-developing structures.

A vivid example of such influence of the medium is famous watermelons and melons supplied in large quantities, in particular, from Central Asia. For many monocultures, not only are the fertile soil and abundant watering important but also the overall temperature background of the medium and its humidity. In conditions of high temperature and insufficient humidity, many productive crops do not yield a good harvest because a huge part of their vital energy is spent fighting heat and dryness.

It's amazing how people managed to get huge watermelons and melons from small berries over time! They need a lot of water and sun rays, as well as a lack of competitors. In primordial nature, these things do not mix; if there is a lot of sun, then there will not be enough water. If the land is not covered with vegetation, then it will dry out quickly in early spring. If there are a lot of weed competitors, they will shade our monoculture and draw in some of the minerals and moisture.

What does a peasant do? He razes the ground, plants melons in one row, and creates an irrigation ditch to organize watering. After each watering, the peasant weeds away, destroying weeds and loosening the soil so that it does not dry out, and carries out selection every year. As a result, thanks to the activities of people, modern varieties of monocultures have appeared, and this can be called the creation of accelerating conditions. As subjects of evolution for the environment, people are no different from melons and watermelons.

Acceleration is also a characteristic for virtual structures, or **virtual reality**. By this notion, I mean consciousness in a broad sense and its special part – science. They are the product of activity of the most complex matter – the brain – and the brain itself is the hegemon in the body. For the development of science and culture, accelerating conditions are necessary; therefore, creative people or scientists who were most intelligent for their time proved themselves more at court or in religious organizations, where appropriate material conditions were created for them. In modern times, the sciences have gone under the wing of the state and business, which provide the best external conditions (environmental conditions) for their development. Acceleration conditions of the medium contributed to the emergence of new subjects of evolution – professional scientists and artists.

Nature promotes acceleration (broadly understood), alternating periods of favorable conditions with climatic crises. This is how it stimulated and accelerated anthropogenesis. Thousands of years ago, people developed and lived in a paradise, but then living hell came upon earth, and humanity has to come up with new ways of existence, which, like in nature, are formed by evolutionary microchanges.

For example, seeing how biological productivity sharply improves in cultural species in the ideal conditions of a subclimate, people themselves began to engage in the breeding of more productive breeds and varieties, achieving success by creating accelerating conditions for selected biological species. Engaging in selection, people accelerate the action of biological evolution in accordance with the theory of Darwin.

On a global scale, the most complex and productive biological species appear due to the gentle level of climate that creates and stabilizes the best living conditions for them.

The Gentle Level of Climate.

As a result of the filling of the land surface with forests and savannas and the formation of inland water bodies, marshes, rivers, and polar glaciers, a very stable and, at the same time, gentle and subtle second level of climate was established on the planet. Or gentle climate, as I call it. We can say that the gentle climate ripens in the bowels of the basic climate; they seem to exist one in another.

The division of the concept of climate into two levels is also connected with the fact that at certain stages of evolution, only one of them plays a leading role, while the other remains in the background. At the onset of the green massif on land, processes of final formation (smoothing) of the relief and the general configuration of land occur and other components of the climatic mechanism of inanimate nature are formed. For example, great river systems moved depositions to the seas, thus expanding the land area over hundreds of millions of years. The Himalayas continue to grow; this affects the climate of the entire planet at a microscopic level.

Nevertheless, the physical evolution of the world is currently obscured by the biological evolution. In conditions of stable and relaxed climate, complex and very productive biological species, such as flowering plants and insects, appeared. Flowering plants gave rise to a new round of green evolution and genetic diversity. The appearance of seed plants makes it possible to cover long distances with the help of wind, and insects are necessary for cross pollination of flowering plants.

The second level of climate must be constantly maintained and permanently improved by biological life, biocenosis, and planetary integrity. At the same time, one must take into account that the living nature is constantly under the threat of aridization because of the 'excessive' power of the sun. I will note that the sun does not shine more strongly, but we strongly contributed to intensifying the effect.

We cannot influence the first level of climate, except for the threat of nuclear war, but we have already thoroughly influenced the second level because the gentle climate is the biosphere itself. Partially destroying and transforming the biosphere, we potentially push back the biological evolution tens of millions of years ago. For now, only potentially.

For us, the first level of climate is conditionally solid because the slow processes of physical evolution are invisible to us; they do not have much significance for the short history of mankind. Over hundreds of millions of years, the physical evolution of the world and the biological life have established the harmony of the climate and its stabilization. Humans violate this harmony; they do it not consciously but out of ignorance or stupidity. We need to blame the anthropogenic factor for today's climate problems – we become the third level of climate formation.

The modern living climatic mechanism of the planet is the result of biological evolution, a function of the gradual increase in thickness and area of vegetation cover on land. The biological life moved upon land bit by bit; it gradually covered the arid and other reactive areas of the land with a green mass, creating sulde. River systems, inland reservoirs, swamps, and glaciers developed.

As already mentioned, the aforementioned tandem of land and water areas is a structural mechanism of climate self-regulation. The water cycle in nature is the basis of the second-level climate mechanism, which depends entirely on the state of the living nature. As a result of the full coverage of the land with woodland, biological evolution has reduced the bare areas of land as much as possible; it has also reduced its reactive heat release to the lower atmosphere to the extreme level. The planet woodland becomes a living air-conditioner. Economically consuming moisture, plants cool themselves and create a microclimate in the near-earth atmosphere.

Planet's sulde changes its quality. Biological evolutionary changes have led to the fact that the hot and hard quality of the land has decreased, and the cool and mobile quality of the world's oceans has increased. At the same time, part of the water from the oceans moved to land, which affected their great ratio. The planet woodland softened the climate; after that, river systems, swamps, lakes, seas, and glaciers appeared – a full-fledged process of the water cycle in nature began.

The water cycle in nature and sulde led to the sustainable development of biological life and the stabilization of climate around the ideal center. The climate stabilizes in a pendulum form, passing the zero point of nature, which science must determine in the form of a figure – a givenness of nature. Now it is characterized by a decrease in the average annual temperature, the frequency and sharpness of its differences, as well as the amplitude (absolute values of the extremes) of these differences.

It is no coincidence that this happened very close to the dew point – such temperature leads to optimal zoning of the climate on land. School physics lessons taught us that the dew point is a ratio of air and humidity temperature, at which evaporation of water into the environment does not occur and water does not condense in the medium. The dew point is a conditional boundary, the neutral entropy of the weather, denoting its equilibrium state.

The average value of annual air temperature on the planet is an indicator of the planet's entropic state as a thermodynamic system; it is a mark of the epochal climate, a step of the conventional scale of the average epoch-making temperature value. Being in the last geological epoch of the physical evolution of the world, we begin to identify the epochs of biological evolution, which we call epochal climatic cycles in the Evolutionary Theory of Climate (ETC).

Epochal Cyclicity of the Climate

Our world is changing in different cycles of time. Before we elaborate on the concept of epochal cyclicity of climate, we formulate the following basic thesis of the Evolutionary Theory of Climate.

Basic thesis of the Evolutionary Theory of Climate No. 4. *The epochal cyclicity of the climate is created by a change in the great ratio of land and the world's oceans (the ratio of their quality) in the planet structure.*

Let us consider the patterns of the emergence of epochal cyclicity of the climate. Biological evolution, which lead to the appearance of sulde, comprising of forests, savannahs, inland waters, swamps, and polar and mountain glaciers, gradually reduces the average annual value of air temperature on the planet, $T°$. The decrease in $T°$ is due to a decrease in the reactive heat released from the land into the surface atmosphere from the effects of solar radiation. The dynamic equilibrium of the second level of climate is a subtle biological and temperature-humidity balance. The rhythmic effects of solar radiation on the planet's surface are mitigated by the complex structure of the planet, which is not static; it is variable and evolves in time.

In the Evolutionary Theory of Climate, I distinguish three levels of the planet's climate cyclicity:
- diurnal fluctuation of air temperature;
- annual fluctuation of air and climate temperature;
- epochal fluctuation of climate.

Diurnal and annual climate cyclicity is associated with the planet's rotation around its axis and the annual rotation of the planet around the Sun. The epochal cyclicity is connected with the dialectic of the general evolution, the essence of which is the development of the world toward complication through the mitigation of the chaos energy in nature.

Diurnal and annual fluctuations initiate the movement and evolutionary development of wildlife despite the frequency and reversibility. In nature, microscopic changes and complications accumulate daily and annually. It is evolution itself. In epochal times, the quantitative accumulation of changes in nature leads to qualitative changes, or phase transitions, when old forms and structures are destroyed and replaced by new ones. Some species disappear and are replaced by others and new alliances of previously independent systems – biocenoses – emerge. Next, we will consider the main causes of fluctuations in $T°$ and the climate.

The history of biological evolution is the history of its struggle with heat, mainly in the middle belt of the planet. Biological evolution gradually increases the woodland area and stabilizes the water cycle in nature. As a result, a gradual decrease in **T°** occurs. Increase of green and water areas and the improvement of their quality leads to the corresponding changes in sulde, as well as an increase in the thermal protection of the planet. The green cover shades the surface of the Earth, protects dense structures of land from direct drying sun rays, and prevents overheating of land. For example, the illumination on the 'bottom' of the rainforest is only 1%.

After complete greening of the planet and achieving the maximum thickness of the forests, the Earth's heat protection system reaches its limit of growth. The rainforests are structured in several tiers, reaching up to 50 m in height, and zones of temperate climate are covered by single-level forests. The elevated valleys in the tropics are not covered with solid forests, but with savannas with groves, because on the elevated parts of the earth, the soil layer forms slowly; it is washed off by rain and blown away by the winds. Due to the thinness of the soil cultural layer on the elevated areas, the wildlife is the most vulnerable during the warming periods on the plateau.

The complication of biological species and their development, the gradual and full-scale coverage of the land by the complex quality of the woodland result in a qualitative change in sulde and reduced **T°**. *Such a period in climate is called the* **epoch of cooling**, *or the ice age.*

Evolutionary processes are accelerated at the final stage of the cooling down period. Huge areas of land are exposed to icing, and the level of the world ocean declines; a significant part of the evaporation of water in the water areas begins to condense at the poles – they build up on the surface of polar glaciers. The modern result of this process is the high level of glaciers in Greenland and Antarctica; it is much higher than the level of the world's oceans.

These processes can be called cold land aridization. Condensation of water occurs mainly in the form of direct freezing, which predominates over evaporation. As a result, some of the water is withdrawn from the cycle in nature and remains in the polar glaciers. On land, dry air prevails like the absolute dryness of air in Antarctica at the present time.

The sky becomes clear and cloudiness decreases. Now the irradiation of the planet intensifies once again, and the 'hour X' is coming for the next phase transition.

During the peak of the cold season, the circulation of water in nature decreases, and, accordingly, the organic nature degrades in temperate climate zones. The planet woodland begins to decrease, and the biological productivity of land is declining. Perhaps, it is in such periods that mammoths died. In the last and accelerated stage of the cold season, the bare land area increases, especially near the shores of the world's oceans.

As we have already decided, on a global scale, the climate depends on the great ratio of land and world's oceans. Therefore, a rapid increase in the area of land, as well as a decrease in water areas and cloudiness leads to a halt in the cooling processes.

Reaching its limit, the reduction of the woodland area due to cold aridization of land is the apogee of the cold season. Figuratively speaking, the climate train stops during an uphill climb and is put on the brake, represented by the planet woodland and the thickness of the polar glaciers.

*Further cooling is stopped by the excess receipt of solar heat and other factors. The reverse transition, or the climate reversion to the previous epochal state, begins. The **epoch of warming** comes again.*

The climatic train of nature, stopped during an uphill climb, releases the brakes very smoothly. First of all, the living nature in temperate zones is restored, and a new climatic zoning gradually takes place. In tens of thousands of years, the planet once again approaches the peak of warming. The specific duration of epochal periods is different because the composition of the evolution 'actors' varies greatly.

Aridization of land occurs in the most vulnerable regions, where modern deserts have developed over time. The processes of desertification do not occur as continuous massifs over large areas: as already proved by paleontologists, large sections of the former natural landscape from woodlands – oases – always remain. The Sahara Desert was an exception to the rules of evolution, because people intervened in the natural course of events. I will give you more details about the history of the Sahara a little later.

Usually, deserts occur on the plateau, in areas with a small layer of soil. If we carefully consider the geographical map, we will see that all the world's deserts emerged on high ground. In the age of warming, the dynamics of **T°** variation once again has an annual microscopic positive tolerance.

What factors could stop the warming processes? First of all, this is the maximum water saturation of land and an increase in the area of the world's oceans, the maximum evaporation of water. The water-green array is necessarily involved in the processes of climate stabilization. In areas with a temperate climate, biological life is restored and gets developed; there is a repeated moistening of the land. A full-fledged cycle of water in nature, which has been narrowed down in the epoch of cooling, is being restored. After maximum moistening and natural landscaping of the land, as well as increasing of the water areas, the era of warming ends and a new climatic period – the epoch of cooling – begins.

There comes a time of continuous cloudiness and cloudy weather, which reduces the access of sunlight to the surface of the planet. If the end of the cold season is marked by constant clear weather and dryness of the atmosphere, like in Antarctica, the end of the warming season involves high humidity and maximum cloudiness.

All the above-mentioned patterns and cyclicity of the climate should be considered only before 'overhuman' comes around. The Evolutionary Theory of Climate does not take into account anthropogenic influence and considers the planet's climate in the natural (pristine) state of nature, which provided its integrity.

The above hypothesis of the reasons for the epochal climate fluctuations is another basic thesis of the ETC.

<u>Basic thesis of the Evolutionary Theory of Climate No. 5.</u> *The physical evolution of the world exists in the form of alternation of geological epochs. Biological evolution exists in the form of alternating climatic epochs of warming and cooling.*

An important note can be made to this thesis: the appearance of humans and their evolutionary development are associated with epochal climate changes, but in the modern world, evolutionary development of nature and climate formation become functions of the human civilization that arose on the planet.

I emphasize that the movement of evolution is a process of eternal continuous search for the entropic balance between the planet and the volatile space as its environment. This is the main characteristic of evolution. Inside the climatic epoch, nature slowly changes microscopically and becomes more complicated. Microscopic changes occur as a result of the effects of daily and annual climate variability, as well as the evolutionary medium as a whole.

Average Temperature of the Planet and the Climate Entropy

The entropic equilibrium is expressed by the stable average temperature of the planet. As already mentioned, this is a kind of conditional dew point for the climate, or neutral entropy of the weather. In the climatic relation, the planet is a thermodynamic system. If there was no annual climate variability, the circumpolar zones would stay in the dew point forever in a state of white nights. Therefore, it is the rhythmic transition of the weather through the dew point that is important for the formation of climate.

The climate of the planet is diverse: warm and humid in the tropical forests, cold and dry in the vast expanses of Antarctica, dry and hot in the deserts, and cold and wet in the taiga. The weather is characterized by three main parameters, namely air temperature, pressure, and humidity; we will call their ratio in the Evolutionary Theory of Climate the **entropy of the weather**. Entropy is not measured by one parameter; it is imaginary and not separately measurable. The quantitative characteristic of entropy for a specific period of time can be represented only in a tabular form.

In addition, we can use a much simplified but basic formula **S (woodland area) /S (land area on the planet) = 1**

We got the **optimum entropy ratio** of the climate, or the ratio of thermal protection of the planet, which is equal to one. 12 thousand years ago, there was a climate in which the green and ice cover spread to the entire middle band on the planet. Simultaneously, at that time, glaciers covered a significant part of Eurasia and part of the world's oceans.

With a total land area of 15 billion hectares, the area of old deserts is 2.0 billion hectares. The area of forests decreased from 7 billion hectares to 3.5 billion hectares; in addition, 2.0 billion hectares of new semi-deserts appeared due to land erosion.

Let's try to calculate the ratio of change in general climate factors in a very rough form.

The areas of degraded lands are as follows:
- old deserts that appeared instead of forests – 2.0 billion hectares
- deforestation in times of industrial revolution – 3.5 billion hectares
- degraded land in the modern era – 2.0 billion hectares

Total: 7.5 billion hectares

S woodland area/S total area = 7.5/15.0 = 0.5

Thus, the planet's **sun protection ratio** is 0.5.

The land areas occupied by humans' artificial world also directly affect entropy in the surface layer of the atmosphere, so the formula needs to be complicated. But we have already gotten a dangerous ratio without it, and we do not know the climate zoning and climate scenario it leads to. But we know the direction of work to remedy the problem.

This formula does not take into account the epochal variability of the climate; for this and other reasons, it cannot be used for specific calculations. Nevertheless, it gives a general quantitative basis for conclusions, as well as a figurative 'cinematic' vision of climate processes.

Biological evolution seeks to achieve the optimal (entropic) state of the planet's climate and create the best conditions for ensuring a high biological productivity of the biosphere by expanding its range and improving habitat conditions. The biological evolution does it through trial and error. Analyzing different options, it achieves its goal through a combination of very complex factors of the climate mechanism and the creation of different climatic zones.

For clarity, let's compare biological evolution with a computer. Computing machines instantly sort through billions of options, selecting the necessary ones or those that correspond to given conditions. Nature checks through these billions of options in the form of field trials, testing them in practice. The evolutionary movement of nature takes place leisurely on a broad front. If the computer conducts a quick theoretical analysis and outputs the result, then biological evolution conducts a slow practical analysis, testing all kinds of options in reality. Then it gives the result, which survives in a medium with varying degrees of volatility.

As a rule, the conduct of practical analysis and experiments by evolution requires a very long time and occurs on a global scale. The distinctive characteristic, the main feature of evolutionary movement in nature is its epochal coverage on time and planetary scale. The adaptation of species and changes in living nature are extremely slow. The comparisons of past and present rates of evolutionary motion are relative, since it is necessary to compare the rate of natural changes in nature and changes occurring at the initiative of a person.

The Main Causes of Climate Destabilization

At the beginning of the chapter, we talked about the achievements of modern scientists in the study of climate problems and the climate as a whole. Now let's try to interpret these achievements based on the Evolutionary Theory of Climate.

First of all, we are talking about the epochality of climate change. We will also keep in mind the concept of the second, or gentle, level of climate and will once again say that the main factors of its stabilization are forests, savannas, and aquatic systems (including glaciers), which together constitute sulde – the natural sun protection of our planet. Third, we will recall that the average temperature of the planet is the equilibrium point of the climate's entropy equilibrium, with the stabilization of which the existing location of the climatic zones on the Earth remains.

Now let's take a look at how we managed to influence the climate. First, we will mentally move to North Africa during aridization of land in this region. Exploring the history of North Africa's climate, scientists made an interesting discovery – it turns out that the climate of the territory, on which the Sahara Desert is currently located, has changed several times over the last 150 thousand years. It used to accommodate tropical forests that, at the peak of the warming epoch, passed into a state of savannas with conservation of tropical forest areas. The savannahs gradually turned into semi-deserts with interspersed savannas residues in them. Semi-deserts, which existed at the beginning of the warming epoch, turned into deserts with frequent sections of oases. At the same time, the total area of these deserts was *no more than a quarter of the area of the modern Sahara*.

During the era of cold weather, the processes were repeated in the opposite direction: the tropical forests and savannahs were restored, and the deserts turned into habitable lands. The epochs of cooling and warming last many thousands of years. During all these transformations, river systems and large lakes never dry up and do not disappear.

In history, such changes were observed directly for geological reasons; sometimes, the climate in the regions depends entirely on them. The essence of such changes is that there is an increase or decrease in landscape level against sea level, but geological processes are always very slow.

The last and complete desertification in North Africa occurred about 5 thousand years ago. The last period of aridization (the period of the climate crisis) spanned 900 years. The peak of the natural period of warming on the whole planet coincided with the increase in the number of people in the region, which caused extreme activity in breeding livestock. Having divided the land, the tribes led a sedentary lifestyle. In savannah-rich areas of North Africa, people began to achieve progress in cattle breeding. The cattle were protected from predators, and people almost destroyed the predators.

People suddenly interfered in the harmony of living nature. Previously, the number of herbivores in North Africa was limited to predator activity, the number of which, in mutual dependence, was limited to the number of herbivores, which also depended on the state of the savannas. There was a dialectical and dynamic balance between fauna and flora.

The profligate breeding of cattle led to overgrazing and rapid denudation of land. Earlier, during natural warming periods, a tropical climate gradually changed to a savannah climate, and the tropical forests, like the seeds of drought-resistant plants, were preserved in the form of small, separated arrays. More precisely, the tropical climate with a stable air temperature of 24-27 °C and high humidity gradually turned into a savanna climate with small temperature changes and moderate humidity.

After the enhancement of people's activity, the rapidly bared land began to dry out intensely and the subsoil inflows of the rivers were disturbed. As a result, land aridization occurred, which led to the complete desertification of the region. This is how the Sahara was formed. In addition, North Africa is a plateau, so its soil layer was very thin and vulnerable to the processes of aridization in the region with a strong level of solar radiation.

The same is happening today in North China. As a result of overgrazing in just 20 years, land aridization in its last stage occurred. Deserts came close to Beijing. In quiet weather, the city suffocates from exhaust gases, and when there is wind, it is accompanied by dust storms. Even car traffic is hampered by poor visibility.

I will add my personal opinion to the above: the emergence of the Sahara Desert led to warming in Europe. I see this as an evolutionary advantage, as the destabilization of the climate on the mainland created a crisis; people were forced to switch to farming. Pain is a cruel joy of being, to paraphrase the name of a well-known documentary.

Now it will be appropriate to formulate another basic thesis of the Evolutionary Theory of Climate.

Basic thesis of the Evolutionary Theory of Climate No. 6. *As a result of slow evolution, the biosphere cannot quickly adapt to sharp changes in environmental conditions for the worse.*

This thesis can be considered one of the main conclusions of the Evolutionary Theory of Climate.

At the end of the twentieth century, scientists in many countries around the world conducted very interesting experiments to study the properties of air. The air was heated in isolated chambers under the direct influence of solar radiation. One such experiment was conducted at Moscow State University in the 1980s. The walls and the bottom of spacious chambers were isolated with effective materials to irradiate only the air, eliminating the influence of the medium. The top of the chambers remained open. This design provided direct access to the sunrays.

In the morning, scientists measured the air temperature outside and inside the chambers: the devices showed the same value, say, + 15 °C. At the end of the day, the measurements were made repeatedly; in the chambers, the temperature was 0.5 °C higher, i.e. +15.5 °C. Due to its weight, cold air stagnated in the chambers without movement, like a poured liquid. At the same time, the outdoor air was heated next to the test chambers in the amplitude usual for Moscow, that is, up to + 25 °C [72].

As a result of the experiments, physicists and climatologists came to the unambiguous conclusion that the atmosphere is incapable of independently heating itself under the influence of direct solar radiation. It was also concluded that, first, the sun warms the surface of the Earth and then the planet, like an oven, transfers the received heat further into the atmosphere.

Physicists' ideas about heat and the processes of heat transfer turned out to be inverted, and the simple mechanical addition of classical physics and the results of the experiments described above led to the emergence of **greenhouse gas climate theory**. A new theory arbitrarily assigned the properties of the atmosphere and a change in its composition as the main cause of global warming. The greenhouse theory pushed the impact of changes in other structures of the planet into the background.

During geography classes, we were convinced that the climate is created and guaranteed by the ratio of land areas to the world's oceans and that it directly depends on the green cover of the planet. This theory unwittingly calms the world community, leads it away from solving a vital problem, and narrows the tasks of human civilization. I would note that the greenhouse gas climate theory is only one of the options, although very popular for interpreting the results of the experiment with direct irradiation of air described above.

Let's consider its physical basis in the aggregate. The sun's rays heat the land; overheated land emits infrared rays; greenhouse gases reflect and delay infrared rays in the near-earth atmosphere. As a result, the greenhouse effect occurs; bare ground produces infrared radiation and molecules of carbon dioxide and methane reflect infrared rays.

It is very difficult to imagine such a physics of processes; it is like trying to catch plankton using fishing nets. Given the small size of the greenhouse gas molecules and their microscopic fraction in the total volume of the atmosphere, it is difficult to imagine the reflection of infrared rays by these molecules.

At least two correct and useful conclusions can be drawn from the greenhouse theory. The first is the principled recognition of the anthropogenic factor of climate change. The second is the recognition of the heating of the Earth's surface by the sun's rays and the emission of infrared radiation back into the atmosphere. However, the greenhouse gas climate theory does not take into account that the exposed lands are a variable factor of the climatic mechanism. The total volume of heat returned to the Earth's surface depends on the area of bare land on the planet. Additionally, it is necessary to take into account the rapidly growing artificial world of people, composed mainly of super-dense materials. Infrared rays are emitted from superheated dense structures from the surface of continents. The amplification of infrared radiation occurs exactly in places where the land is bare and not protected by woodland from direct sunrays.

In any locality, the climate and weather are affected by large cities and the amount of buildings and communications. A completely new and very strong climatic factor has occurred. The emergence of cities made of concrete, metal, and glass should be assessed as a change in the reactive properties of land as a geological change. The quality of land changes for the worse. For example, in cities, the average outside air temperature is always 3-5 °C above normal temperature. In this context, the usual temperature is the average air temperature in the region in past times when there were no modern-type cities, the pristine nature was preserved, and the amount of water and woodland was enough. The increased temperature of the city affects the temperature of the surrounding area, increasing it by 1-2 °C compared to normal temperature.

Increasingly hot weather on local parts of the planet makes the planet's climate diverse. Bare and dehydrated lands, new arid zones and cities heat the atmosphere in smaller areas in certain regions. The number of such sites on Earth grows rapidly as the number of cities increases – as of today, it is more than two million.

In addition, the climate is microscopically influenced by settlements, farms, and other agricultural lands. The air temperature around the railways and highways is always increased; iron roofs of houses intensively influence the air temperature in the cities. Each of us feels the heat, but not all pay attention to this in the scientific sense.

Try to sit under an iron canopy on a summer day, and then relax under a wooden one. You will feel a colossal difference; it is as if you go first to hell and then to heaven. If you are outdoors far from the city, then, when crossing the nearest highway, you will always feel a warm air strip – you feel it even though you may not scientifically comprehend it. Today, the air temperature of every house, structure, and chimney is higher than it was 100 years ago.

People began to observe weather using various instruments in the nineteenth century, however, they were unsystematic. Meteorological services of different countries have observed weather according to uniform standards only since the beginning of the 20th century. Therefore, we build on data on the planet's climate over a hundred years.

Earth's rotation and other causes mix weather conditions; as a result, the average temperature of the planet rises against the data of a hundred years ago. This begins to negatively affect even the regions still undisturbed by man. The general increased climatic background affects the state of the entire biosphere – tropical, deciduous, and mixed forests, taiga, and polar and mountain glaciers. In North Africa, with which we began to address the issue of climate destabilization, the Sahara annually extends for 6-10 km to the south; in proportion to this, degradation of the tropical forests of Central Africa occurs.

Due to the increase in the planet's average temperature, the groundwater level, air humidity, and the warm period of the year are decreasing everywhere. For example, bark beetles manage to give offspring 2-3 times per season instead of the usual one; as a result, the trees affected by the beetles fall sick. Similar phenomena occur in North America, in Europe, and in the forests of Asia. Due to the dryness and diseases, all wildlife is decrepit; forest fires burn around the world every year.

Previously, because of high humidity in the taiga forests of Siberia, it was difficult to find dry brushwood to start a fire; matches would soak. In just 50 years, the situation radically changed because of the aridization of land. Ground water in the taiga is reduced by 4 meters, and the level of Lake Baikal is also reduced. Now the forests burn even in protected areas, which saw no direct human intervention.

Many climatologists look for the cause of global warming in natural epochal processes; some of them believe it is the atmosphere that has to do with the increase in the proportion of greenhouse gases. I believe that the current structural changes on the surface of the Earth have strongly affected the planet's climate. All side effects on the ground turned into one big climate change.

At the Paris Climate Forum 2015, scientists and leaders of high-ranking states declared that the current warming can be tolerated (in other words, our civilization can afford it) at 2 °C until the end of the 21st century. It means that the previous change of 0.85 °C over 100 years can grow up to 2 °C. Such an assumption was born after some studies showed that on the peaks of past eras of warming, the average temperature was indeed very high.

Why is this happening? What is special about the current situation? Is the story with North Africa repeating itself? My answer is yes. It is repeated on a global scale all over the world. Climate destabilization is due to significant changes in the planet structures, that is, crises have a solid character and a fundamental material basis. First and foremost, the plant world cannot adapt to rapid structural disturbances and global warming. Structural changes have also occurred on the surface of the Earth: the green cover of the planet has decreased by half. An additional factor affecting the climate crisis is the artificial world, or the anthropogenic factor.

In reality, it was enough to raise the average temperature of the planet by only 0.85 °C in order to initiate the beginning of global processes of aridization of the entire land and accelerated melting of polar glaciers. An increase in the average temperature by 0.85 °C may seem insignificant; in fact, it leads to huge changes on the surface of the Earth.

There is a serious tension in the thermodynamic system of the planet. Many scientists rightly consider it to be an integral and living organism, and today, we have changed the normal healthy temperature of the Earth. The last period of the natural warming era in a particular region – namely, in North Africa – lasted 900 years. Above, we considered how a flowering landscape with groves and savannas, powerful river systems and lakes turned into a desert during this period. Aridization of land in the region was occurring slowly, with microscopic steps. According to experts, the speed of the current processes of warming and aridization is an order of magnitude higher than in past warming epochs.

If the processes of warming took an evolutionary path, taking place very slowly, then the living nature could adapt to them. There is a serious concern that the changes take an exponential character because the melting of glaciers is accelerating from year to year, and climatic fires become a full-scale and systemic threat.

The aridization of land, as well as the decrepitude of the biosphere, take place on a global scale. At present, there is not a single corner of the planet that climate change has not touched.

It is safe to say that the structural changes on the planet have not yet fully affected the climate; an increase of 0.85 °C is only a visible part of the iceberg. The real average temperature of the planet will come to light when the processes of searching for entropy equilibrium are complete, which is very undesirable. I would compare our planet to a patient who has suffered a high degree of skin burns. The damage received has not completely affected the future state of the patient; it is impossible to accurately predict what it will be; but the main thing is absolutely clear: the patient may not survive without active medical care.

Let me explain with an example what searching for entropy balance in a thermodynamic system is. Suppose you install a heating device in a cold room. The air temperature inside the room gradually rises and reaches its maximum +18 °C. You install another heater and wait for a further increase in temperature. Today, we have the same expectation: we have increased the reactive quality of land and do not know what it will lead to.

The balance in nature has been disturbed and we do not yet know when and how the degradation processes will stop. Facts are an inexorable thing; they suggest that the processes of land drying are only intensifying. Aridization of land disrupts the balance of evaporation and condensation of water in a particular climatic region or on a local site. There is an excess of the volume of water vapor over precipitation, which leads to further desertification of land and complete drying of inland water bodies and river systems.

The rise in temperature on local sites affects the overall climate background; this phenomenon is also undeniable, as this is the arithmetic mean. The melting of polar glaciers and permafrost accelerates. Warming leads to forest fires around the world. As a result, the degradation of wildlife takes place, which, in turn, additionally 'promotes' the climate crisis.

Deprived of its green cover, the bare ground is not an intermediate, but the main factor of warming. The loss of the planet's sun protection and the addition of the artificial world reaction lead to the aridization of lands and the disruption of the water cycle in nature. Currently, emergency zoning of climatic zones has begun. All of this adds up to the degradation of living nature.

Over the past 200 years, people destroyed half the world's forests; they cut down trees on the territory of 3.5 billion hectares. In addition to this, we have subjected approximately 1.5 billion hectares of land to wind and salt erosion and built an artificial world of dense materials. In terms of intensity of infrared radiation, it is more reactive than the bare ground.

Direct emission of heat by human economic activity, which is approximately 2% of the volume of solar energy received, may well affect the shifting of the epochal climate's vector in the balance of nature. The decomposition of billions of tons of garbage and their latent burning also affect the climate. Any decomposition, including the burning of hydrocarbons, leads to climate imbalance.

Basic thesis of the Evolutionary Theory of Climate No. 7. *The phenomenon of infrared radiation is the release of heat by dense structures of the planet directly into the near-Earth atmosphere. Infrared radiation creates heat. Unlike solar radiation, infrared radiation leads to a direct emission of amions into the atmosphere.*

This thesis is, perhaps, the main and most important thesis of the Evolutionary Theory of Climate, as it contains the essence of a new scientific approach to the relationship between man and nature.

Solar radiation carries kinetic energy (if there is a small number of amions), and infrared radiation is accompanied by a direct emission of amions and their 'clots' – neutrons.

As is our custom, I will try to explain this thesis at the level of a domestic example. It will be more appropriate to do this in the next section, which I will start off with this example.

Vortices and Other Phenomena in the Lower Atmosphere

In the summer period, the walls of houses heated over a light day accumulate heat. At night, some of the heat is released into the living quarters, so in hot countries, people prefer to rest outdoors at night. It is not unreasonable to assume that the land heated during the day emits thermal radiation into the surface atmosphere. What is the role and special meaning of its condition? Earlier, we talked about the immobility of the lower layers of the atmosphere and the fact that the Coriolis force does not apply there. The lower layers of the amionosphere are dominated by the gravity forces of the Earth; this is the essence of the nature of anticyclones.

Imagine that you are driving a car at a speed of 120 km/h. Extending your arm outside, you feel a counter-flow of air, which has exactly the same speed. Then you decide to ride a super-fast carousel. Let's assume that it rotates so quickly that the linear speed of the cage is also equal to 120 km/h. Thrusting your hand through the cabin window, you will once again feel the oncoming movement of air with the same speed.

Now imagine that our planet is a large space carousel. Sitting at home, you are moving at a speed of 1500 km/h. But you will not feel the same speed of the headwind putting your hand out of the window. The thing is, all objects and air are attracted to the surface by gravity, and the carrier of these invisible forces is the amionosphere. Amions, air molecules, and all objects and living beings on the surface of the planet, including us, fly by inertia along with the Earth.

Such synchronism occurs only in the lower layers of the atmosphere. In the troposphere and above, the atmosphere lags behind the planet's rotation, so the Coriolis force appears there. Accordingly, there are constant satellite vortexes in the high layers. In the ecliptic plane, the atmospheric vortexes arise from the interaction of the forces of gravity and the forces of kinetics of amions.

Asynchronous combination of these two forces leads to the appearance of torque and the Coriolis force. These vortexes constantly and continuously occur in the atmosphere, and their axis of rotation is almost parallel to the axis of rotation of the planet; more precisely, they are somewhat inclined to the poles – to the North Pole in the Northern Hemisphere and to the South Pole in the Southern Hemisphere.

Our planet rotates around its axis. The atmosphere (the amionosphere) lags behind the angular speed of the Earth; each of its layer brakes separately. For the above reasons, it is mixed continuously. The air seems to shift relative to the surface of the planet and, at the same time, gets shoveled. An important fact here is that the Coriolis force practically does not affect the lowest layer of the atmosphere. The main role in the mixing of air is played by other physical laws.

Due to the limitation of the action of the Coriolis force on the surface layers of the atmosphere, there is a 'stagnation of the climate' in its lower layers – a phenomenon known to us as anticyclones. As a result, the vortexes familiar to us do not affect the troposphere with the help of the Coriolis force. It is very good, as this becomes a factor in climate stabilization, but does not mean that the atmosphere of the lower layers does not move relative to the Earth. Its displacement occurs without the formation of vortexes in the ecliptic plane. Basically, air mixing occurs in the regime of soft (non-vortex) diffusion processes.

To understand what vortexes are in the ecliptic plane, we need to compare them with the vortexes, or cyclones, which occur in the lower layers of the atmosphere. Natural vortexes rotate in a plane parallel to the Earth's surface. Their axis of rotation is directed vertically with respect to the horizon, and the vortexes move the air between the climatic zones. For example, they can move the cold air of the Arctic Ocean to Central Asia, which, by the way, happens very often. In Coriolis vortexes, the axis of rotation is parallel to the axis of rotation of the planet, so the air moves these vortexes from the bottom up. Most importantly, these processes never stop.

The vortexes that we see – cyclones – appear in the lower layers of the atmosphere. They occur in the plane of the Earth's surface and their axis of rotation is directed toward the center of the Earth. Similar vortexes occur in the ocean waters on a huge scale. It is clear that the Coriolis force is not involved in these vortexes since it can arise only in the ecliptic plane.

In the chapter, "The Formation of the Solar System," we established that two forces of amions are required for the appearance of vortexes: forces of attraction and forces of inertia. The only question is: why is the axis of rotation of the vortexes and large cyclones in the lower layers of the atmosphere perpendicular to the surface of the planet? It turns out that the answer is very simple: vortexes occur in a plane where the gravity forces of the Earth do not work.

Now consider the chart in Figures 34 and 35, where the abscissa axis is the density of the amionosphere, and the ordinate axis is the height of the amionospheric layers. The values are completely identical to those of the atmosphere. As you can see, the chart has the form of a quadratic function, where the density decreases exponentially as the height of the amionospheric layers increases. In normal quiet times, the chart is complied with.

When is it violated? When there are several cloudy days and it is cold; then the sky becomes clear, and the lower layers of the atmosphere suddenly heat up. I emphasize that only the lower layer of the atmosphere will heat up sharply. In our chart, there is a line; let's call it a yield line. The shaded part of the chart is surplus amions, as we called them earlier, 'unemployed' amions; they do not 'serve' the planet's gravitational network. It is these 'unemployed' amions that initiate the vertical vortexes in the atmosphere. In recent years, such vortexes – cyclones – occur more often, and their strength increases exponentially. We will further refer to this chart in the Evolutionary Theory of Climate as **dissonant schedule**.

It seems to me, based on what you have already read, that you should have a fairly complete idea of the new Evolutionary Theory of Climate. In this chapter, its main points are covered in sufficient detail for any reader, regardless of their scientific background. They are simple and accessible for understanding and quick uptake of content.

Deepening into the details of any of the above questions or topics, we will inevitably have to speak the language of special professional terms and mathematical formulas. Therefore, in conclusion of the chapter, I suggest to simply accumulate in the final section the main theses of the Evolutionary Theory of Climate, which were previously specially identified in the text and numbered. At the moment, there are seven of them.

Preliminary Results: 7 Basic Theses of the Evolutionary Theory of Climate

<u>Basic thesis of the Evolutionary Theory of Climate No. 1.</u> Climate is the achievement of world evolution. The modern soft and stable climate was created by nature over billions of years of physical, and then biological evolution. Gradual mitigation of climate is a necessary condition for evolutionary progress.

<u>Basic thesis of the Evolutionary Theory of Climate No. 2.</u> There are two levels of climate: basic level, which is a result of physical evolution and gentle level, a result of biological evolution.

Basic thesis of the Evolutionary Theory of Climate No. 3. The planet Earth initially gets an excessive amount of heat from the Sun. It is constantly under the threat of global warming.

Basic thesis of the Evolutionary Theory of Climate No. 4. The epochal cyclicity of the climate is created by a change in the great ratio of land and the world's oceans in the planet structure.

Basic thesis of the Evolutionary Theory of Climate No. 5. The physical evolution of the world exists in the form of alternation of geological epochs. Biological evolution exists in the form of alternating climatic epochs of warming and cooling.

Basic thesis of the Evolutionary Theory of Climate No. 6. As a result of slow evolution, the biosphere cannot quickly adapt to sharp changes in environmental conditions for the worse.

Basic thesis of the Evolutionary Theory of Climate No. 7. The phenomenon of infrared radiation is the release of heat by dense structures of the planet directly into the near-Earth atmosphere. Infrared radiation creates heat. Unlike solar radiation, infrared radiation leads to a direct emission of amions into the atmosphere.

CONCLUSION

Evolution of Consciousness

The solar system is unified and organic. By its organic nature, I call it structuredness in the form of flexible element-by-element systemic links in the overall structure. It is understood that physical structures are self-controlled and adapt in the medium to preserve their complexity. The evolutionary development of the physical world occurs along an expanding spiral. The inanimate world passes through the same mandatory stages of self-development that have already been recognized in biological evolution.

Biological forms begin with viruses and other microorganisms: they have 'skin,' or crust, separating them from the medium; there is a nucleus and a body, or protoplasm from a biological 'broth.' This is a permanent reserve of 'spare parts at hand' for self-recovering of the nucleus. Microorganisms form unions; the union of cells already has a common 'skin,' a common organism. At the same time, cells (microorganisms) specialize; some make up 'skin,' others form digestive system, and so on. Later, a swarm (a flock) appears. In this case, it is no longer the union of cells, but individuals, where one of them becomes the main one and others serve it.

Then the whole scheme is repeated in society.

The above processes are repeated in the physical world. Moreover, it is in it that the principles of self-complication of matter are laid. In physics, all material structures have a center, periphery, and crust. They also metabolize with the medium and adapt to it. Complex structures are even more complicated; they create unions of atoms – substances. We come to the conclusion that life begins at the level of atoms; biological life is a repetition of the same, but already at the highest part of the developmental spiral.

The development of consciousness is common to them. We can see its higher level on the highest part of the philosophical (evolutionary) spiral. I call the above principles **canons**.

An example. The substance in the form of a metal plate reacts to solar radiation. These reactions are repeated always and forever, and the plate acts like a natural machine. The atoms, and then the substance (the union of atoms) have the simplest consciousness – stable algorithms of behavior and adaptation.

At the same time, it is necessary to distinguish between the algorithmic atomic consciousness and the consciousness of matter, which is composed of a set of consciousness of individual atoms, where each atom has 'decided' that it is more profitable for it to have company. With sudden changes in the medium, the existence of a union becomes impossible; the atoms make a decision about surviving apart. Perhaps, you thought that this happens in the conditions of absolute cold. No, this almost always happens in the conditions of warming (favorable conditions).

We discover that the development of consciousness has always been and continues to be in the focus of evolution; it is common to all kinds of complex matter. I dare to call it soul, or **virtual reality**. We have shown above that a primitive and simultaneously fundamental consciousness is first born in atoms.

The evolution of consciousness begins with the appearance of the first complex physical structures; atoms and substances composed of them are primarily a natural machine. The automatism of atomic consciousness is inherent in the structure itself – the atom does not have neurons. Therefore, atoms and substances always react unequivocally to external influences and changes in the environment. That's where the constancy of physical laws, their repeatability, and uniqueness begin.

The principle of certainty of quantum physics should be clarified as follows: there are always contradictions within the atom, but its behavior in the external environment is unambiguous. The atom is super stable because it muffles all its contradictions inside itself through **implosion**. Implosion produces emanation (emission) of amions into the external medium, and the center of the atom produces assimilation of amions from the outside and compensates for the loss of matter, complying with the principle of completeness. I call this phenomenon the metabolism between the atom and the medium, or the breathing of the atom. The atom breathes.

Since the emergence of biological forms of the complication of matter, the consciousness of biological species turns from simple atomic machines into a 'thinking device.' To respond to external influences, individuals have many options for adaptation and behavior. Comparing biological species with atoms, we need not miss such an opportunity and point out the ambiguous behavior of the former in the environment. In practice, the principle of uncertainty begins with biological devices; this is what makes them fundamentally different from inanimate nature.

Aristotle writes the following in the first lines of his Metaphysics: "All people naturally aspire to knowledge. Proof is the attraction to sensory perceptions, because regardless of whether they are useful or not, they are valued for their own sake; the visual perception is most appreciated, for we prefer the vision to all else. Vision contributes to our knowledge more than all other senses." [2]

Why did Aristotle talk primarily about knowledge as the main property of man? We note the following points in evolution. Nature was engaged in more variety of species and perfection of biological forms. Why is man chosen as the last object and subject of evolution? Because he turned out to possess perfect consciousness.

Jumping from branch to branch, the monkey develops spatial consciousness; it is its advantage, which predators in the flat world do not have. As a result, developed hands and spatial orientation (the ability to balance with the body) gave hominids the advantage of bipedalism. In the tropical forest, in the conditions of excess food, the monkeys had an all-season estrus, which created a complex family connection. In addition, the favorable conditions of paradise provided a lot of time for games, and the game is an imitation of life, the development of abstract thinking. In the process of endless games, speech appeared as a result of self-development of abstract thinking.

Human evolution is rather the development of content than form. The content of a person is the self-development of their consciousness in society. Each person develops their own consciousness themselves through the knowledge of the surrounding world; the essence of a person is an indispensable and continuous desire for knowledge. The individual is simmering in the vat of society, and the integration of individual consciousness is formed into the public consciousness.

What are the fundamental properties that distinguish the non-living nature from the living nature, and how do they flow from one another? Let me put it this way: how does the relationship between complex matter (complex material structures) and the external environment develop, and where does the world system – the hierarchy of nature – appear and develop? If we can answer these questions, then we will approach the solution of the basic question of philosophy, but we will talk about this later.

The latest achievement of the whole evolutionary movement is the human consciousness. In the last centuries, its development has accelerated dramatically. The world is experiencing strong development of science; it is followed by the formation of **world public consciousness** at an increased pace. This is a phenomenon and virtual reality, and this is exactly that. Today, it turns from consciousness for people into the planetary consciousness and becomes the consciousness of the whole planet. We, people, begin to account for the interests of the surrounding nature (the planet), recognizing ourselves as its organic part.

One can say with certainty that our planet is beginning to have a self-consciousness. The self-consciousness of the planet exists in the consciousness of specific people, the formal and informal elite, and is transformed into a public consciousness.

At present, a paradigm for the development of mankind of the second order is being formed. An undeniable fact is that today, thousands of scientists around the world are concerned about the global problems of the era. This concludes the period or the era of **spontaneous development of mankind**. For thousands of years, humans have developed under the auspices of biological evolution; at the same time, we have always used the old evolutionary method of nature, its universal method of trial and error.

Countless trials and errors are prompted by the active part of our consciousness, and a multitude of mistakes and rare successes lead to the formation of a systemic consciousness. Our consciousness has two parts: active principle and organizational principle, or system consciousness, which, as it turned out, is formed integrally only in the form of planetary consciousness. It works on the principle of "all or nothing." To live well, today, we need holistic, systemic knowledge. Otherwise, it is impossible.

What kind of knowledge is it? In the 17th century, the rapid development of the natural sciences began. In their days, noting such trends, Francis Bacon and Rene Descartes formulated and declared a conscious paradigm of human development. They saw the ultimate goal of knowledge in the domination of man over the forces of nature, in the discovery and invention of technical means, in the knowledge of causes and actions, and in the improvement of human nature. Up to the present time, the paradigm of the development of mankind meant that it is necessary to take more from nature with the least effort.

The people's desire to create a special microclimate for themselves is equally important; for this purpose, they continuously build and expand the artificial world. People create civilization, their isolated 'world' inside the big world. We have achieved fantastic 'success' in the development of civilization, however, we have paid very dearly for this pleasure, violating the structural equilibrium of nature and exhausting the resources of its sustainability.

The old paradigm of human development did not take into account the interests of 'little brothers.' Globally, we have never thought about the limits of our expansion into the primeval nature. In fact, our 'little brothers' are much older than us; that's why they are much smarter than us. In this case, the concept of mind is used in the sense that affects the adaptability of species.

More precisely, in the geosphere and the biosphere, there is a virtual reality, or planetary consciousness, which consists of local consciousness of all kinds of biological and physical matter. Let us clarify which consciousness of nature and the entire planetary system we are talking about. Usually, we confuse this concept with self-consciousness, and the planet does not have it yet; we ourselves must become the self-consciousness of the planet, as this is our evolutionary purpose.

The values of things are determined by comparing them to their likes. Let's compare our consciousness with the planetary one. We have the consciousness of the whole organism, like the consciousness of the planet. In this respect, we are a holistic and self-sufficient structure. All of our body organs work smoothly like a clock: if we are cold, we start to tremble, and if it is hot, we automatically begin to sweat. In our body, there are 4 kg of bacteria, some of which are employees of our internal chemical plant – the digestive system. Another part of the bacteria is engaged in supplying the whole body with oxygen, and one more part protects our body from alien microorganisms and everything else. If our skin – our materialized border with the whole world – is pierced or injured, a whole army of lymphocytes will materialize in this place, declaring war on bacteria. Our whole body is controlled by our mind, but it is still not self-consciousness. Self-consciousness comes later; it does not manage the whole body, as it is not its function.

As a result, varieties of system and local consciousness in nature constitute something common, complex, and harmonious – a virtual planetary consciousness. Having such a perfect consciousness – adaptability, harmony, and stability – why did nature begin to deal with human evolution? A quite possible and logical answer is that it [the planet] lacked self-consciousness, and now a critical consciousness is required for the continuation of a new stage of evolution. Nature does not do anything unless it seems appropriate.

Self-awareness and awareness are primarily a critical consciousness. Any consciousness will be complete and self-sufficient when it is planetary and when it has a critical self-awareness.

Let's formulate the problem of the whole book, denoting the attractor of its development. The predetermined goal of the whole evolution, or an attractor, calling and drawing us from the future, is the completeness and self-sufficiency of our consciousness. This is where the principle of "all or nothing" is applied. Half consciousness means that we have no consciousness: as they say, you cannot be half pregnant.

We acknowledge that all our contemporary problems come from the imperfection of our consciousness, its incompleteness. Studying the book, we need to know what it is written for, so we want to replenish consciousness and our knowledge of nature to the stage of self-sufficiency. In each paragraph and section of the book, my purpose is to fill the gaps in science and find alternative ways of its development. This is the only way to achieve the perfection of man and society.

In our era, the main task of mankind is to bring self-awareness to the planetary consciousness (the consciousness of the planet), which means understanding that the planet and us make up a single whole. We must feel the state of all processes and things in the surrounding world and realize them at the scientific level. Further, already at the level of critical self-awareness, we must assume the function of the virtual consciousness of the entire planet. We should see ourselves in the form of the planet's mind and recognize it as part of a coherent and organic system.

The noosphere is not something separate from everything else; it is the virtual quality of the planet, the result of self-development and the complication of matter over billions of years.

Phenomenon of Awareness and its Role in the Evolution of Consciousness

In comparison with the integral consciousness of the planet, people have a predominantly one-sided consciousness; we have only developed its active part. The boy sitting on the elephant's neck easily controls the animal; his consciousness is locally stronger than that of the elephant. But all subjects of evolution in nature (the biosphere and the geosphere) have a harmonized planetary consciousness, a certain social contract of agreement on integrated development, including the physics of the planet, between species and individuals.

People need to overcome the psychological frontier in the development of their minds and adaptability, dissolve the demarcation of isolation between consciousness and nature. We again need to go back and enter into an alliance with all nature; until then, we will be alien in this world.

In our biological basis, we are a separate species different from others by the presence of consciousness and self-awareness. That is, we comprehend that we are. In an anthill, there is a central individual and specialization among ants: some ants are guards, others are simple workers; an anthill has a collective consciousness and individual separate groups of ants have a specialized consciousness.

But these 'specialists' do not have individuality; in contrast to them, people do have it; that is, people distinguish themselves from others; each of us has a self-awareness. Awareness is a new phenomenon inside the phenomenon of consciousness and a new stage in its development and the development of public consciousness. Now the stages of the development of self-consciousness within the consciousness can be seen; some have more of this secondary phenomenon and some have less.

Self-awareness becomes self-love (ego). We do our best to love ourselves; as a result, a new phenomenon becomes a great stimulus in the development of people, contributing to the development of general consciousness and science. It is no secret that any scientist today dreams of becoming a Nobel Prize laureate.

Where does the phenomenon of self-consciousness move in its development? Previously, the phenomenon of consciousness (personality) in nature did not exist; perhaps, it existed in another form? The philosophical side of the issue: we must formulate one of the principles (canons) of evolution. But first, we will answer the previously framed question: what is the difference between a living being and inanimate nature?

The development of physical nature is systemic and all-encompassing. If it is hot in the environment, all the atoms experience this heat together. As a result, all the atoms of matter come to a new adaptive position. The physical part of the planet is a single atom. It lives in the Solar System and adapts itself comprehensively to the impacts from outer space. Atoms, or unions of atoms (bodies), never 'think' about some of their isolated lives. Our planet without its biological part has only a complex consciousness; accordingly, in the physical world, there is absolutely no 'individuality.'

Biological forms begin to have individuality and individual consciousness. Each individual microorganism becomes an autonomous device; it has a body protected by crust; there is a center that stores that very program of individuality – the DNA code. Simply put, every living organism is an independent world that survives apart and at the same time strongly depends on the state of the physical part of the world.

Why did individuality appear and why is it needed for biological species?

Individuality appeared to preserve the structure of a very complex matter, for which special environment conditions are required; a 'broth' of complex nutrients and conditions for self-development, and restoration of such 'broths' is always needed. I will not go into great depth on this – much has been said about this in biology books – so I will only refer to the philosophical side of the issue from the **general theory of evolution**'s point of view. For example, to preserve the constant self-recovery and replication of the super-complex DNA-RNA molecule, a holistic biological (live) mechanism is needed; the rest of the cell body is an artificial medium for DNA.

Let me reiterate my guesses: atoms are so stable that they do not need special conditions, whereas for complex organic molecules, they are an absolute necessity. Atoms do not need a special medium because their 'oxygen' – amions – is always at hand. A cell is a detached world. DNA for 'breathing', or replication, requires a special medium – a 'broth' of complex organic molecules.

I came to this understanding of very complex biological things after mastering the basic rules of evolutionary logic.

In the world evolution, the following hierarchy of complexity is observed.

- Atoms: a primitive automatic consciousness.
- Biological species: consciousness + individuality.
- People: consciousness + individuality + self-awareness.
- Planet: planetary consciousness + individuality + **self-awareness.**

In the final version of the fourth stage, we humans should first realize the missing part (individuality and self-awareness of the planet) in a scientific way, and then create it.

Before humans, biological forms did not have self-consciousness; it is this quality that distinguishes us from the animal world. People have a physical structure, consciousness (individuality), as well as **self-awareness.**

Essentially, self-awareness is a critical consciousness. Consciousness knows, or realizes that it exists; then it can and is beginning to critically evaluate itself. In addition, it can influence the level of its development.

Self-awareness appears together with the phenomenon of abstract way of thinking. As a result of this double spontaneous generation, a kind of virtual reality appears, an endless world of fantasy is born in the human brain. The existence of a virtual world, or virtual reality, is an indisputable fact. As a result, people are able to predict the future in different variations; in their thoughts, they are able to create an already fictional world and its fictional rules.

As a result, a person begins to live between wakefulness and dreams. Aristotle called it in one word - metaphysics, or "before physics." I would correct the formula of metaphysics as "before and after physics." Such a formula would mean that metaphysics is everything that is outside of physics and surrounds it. Physics is the visible and tangible reality of the world, and metaphysics are our fantasies, which include both the invisible reality and our fantasies about them.

Giving rise to self-awareness and self-love of people, nature made a great trick; namely, it forced each of us to engage in self-development. Self-awareness adjoins (gets attached) to the already perfected ideal consciousness of biological individuals. Nature believes that it reaches a new level in the development of consciousness, giving it a stimulus of self-consciousness, a double-edged thing. This is followed by self-love – the ego.

On the one hand, the appearance and dominance of egoism in the ontological essence of people and their lives is good (positive in evolution). On the other hand, when it comes to the sustainability of the entire planetary system, this phenomenon becomes a threat to the entire nature.

For this reason, an acute problem in the relationship between people and nature, their Alma mater, is brewing in the whole world today. Alma mater is Latin for nourishing mother. Philosophy, or more precisely, evolutionary ethics, is faced with a simple and, at the same time, the most important question: we need to deprive our consciousness of its desire for monogamy, the desire to serve only its ego. We must deprive our consciousness surgically of its 'maidenly innocence' (the desire for monogamy) as early as in childhood. First of all, it must serve society and nature.

At our stage of development, surgical intervention is a necessity because we are in time trouble, in a situation of survival. Later, after the crisis, it will be possible to continue the peaceful stage of evolution. We are currently in a situation of phase transition. This phase is special; it is a phase squared and science has no words to describe it.

I would very much like to call it the **adal phase of evolution** in honor of my friend Adal. With only high school education, he immediately began to understand me. Scientists could not believe me, but he did, and this is a phenomenon. He has a rare original (not clouded by civilization) consciousness, and this allowed me to reach out to him. Without his selfless assistance, this book would not have been published. I would have burned up my own thoughts; an explosion would have happened inside my brain and break my psyche. We called this phenomenon implosion.

Why do I compare Adal with the features of the present phase of evolution? There is a direct analogy (parallel): I told him about my visions, he made a conscious decision and began to help me. His faith tripled my strength, and I was once again convinced of my rightness. Adal is the first person in the world who believed me.

The peculiarity of the present moment (phase) of evolution is that now we must make a conscious decision about our development; this will be a decisive turn in our minds. From the stage of spontaneous development, we will move to the next conscious stage.

We, like ants, must first of all think about our 'anthill,' represented by society and all of nature. This is our return to the old and, at the same time, new state in a philosophical spiral. We once served only the pack, then the race, and then our ego. Now we are moving into a new expanded level of deeds – serving the whole planet.

We all love ourselves and try to create better living conditions, and this requires taking more and more resources from nature. At the same time, people are getting more and more; over 200 years of the industrial era, the world's population has increased seven-fold: in 1800, there were 906 million people, whereas in 2000, there were 6,500 million people. The level of consumption of the average person in Western countries and all over the world sharply increased. As noted by a famous American writer, the standard of living of an average American is much higher today than the standard of living of medieval kings. The aristocrats did not know what a bathhouse was and bathed in a barrel; neither did they know what underwear was, and so on.

Gradually, many people come to a new level of awareness of planetary problems. One of the main points in the development of our consciousness is that these trends are being hampered by science. In matters of climate, modern philosophy has no support from physics. For this reason, I started working on **new beginnings of physics**, develop a new definition of heat in the atmosphere, and so on. Now I feel – with all my sense organs and a new scientific consciousness – that the trees are our mediated skin, and the green cover of the planet protects our world from the searing rays of the sun!

In the twentieth century, after the end of World War II, rapid economic development began in most countries, and people's pressure on nature increased. Scientists were first to sound the alarm, as they noticed that climate change had begun on the planet. Then, opposition of a new kind appeared in the West. It should be noted that such people appeared for the first time in the history of all mankind. A new 'breed' of people who think this special way is a completely new phenomenon in the whole history, except for rare philosophers from the distant past. It was followed by various social movements; in addition, there are many people around the world who sympathize with the so-called 'tree-huggers.' Ecologists, 'tree-huggers,' anti-globalists and representatives of other movements are gradually changing the public consciousness in the world. In the global world today, the ideas of evolutionism have taken the place of socialist ideas.

It is an indisputable fact that real contradictions between the interests of nature and the interests of people have surfaced in the world; the old paradigm of our development, which included "taking more and more from nature," has begun to fail us.

Let's look for and formulate the idea and purpose of the whole world evolution – the paradigm of the development of nature in general. It works out like this: nature devotes its resources to the development of the world consciousness, creating more and more perfect biological forms and providing them with an improved consciousness and a better level of adaptability.

Creating a human, world evolution reached its biological perfection; however, it limited itself to developing only the external form of people, 'naked monkeys.'

Then, nature pursues its main goal – the accelerated development of social consciousness contributes to this in every way. The general evolution of nature and its achievements create a favorable environment, and general principles (canons) realize the development of people's consciousness through the creation of a perfect society and moral and other norms.

The development of mankind occurs within the framework of the above general paradigm of nature. For us, it consists of two parts: the development of human consciousness (childhood) and its transformation into planetary consciousness. The human consciousness – including the development of science and technology and the creation of subtle technologies – is the focus of the general evolution. At the same time, it is the subject of its impact. As of today, almost all developed countries have chosen this direction. A special and conscious goal of humanity at this stage of development is reducing pressure on the pristine nature.

However, all this was not enough. Today, the general theory of evolution and complete knowledge is the breath of life to us.

The Beginning of the Conscious Stage in Evolution

If we call the early stage of development of mankind the childhood period in its evolution and the spontaneous development of consciousness and science, then the mature period of the active and conscious development of mankind that positively influences further development of the planet must come next. The next stage of development should become a stage of the conscious period of the entire evolution of the world. Under the threat of global problems, we are forced to conclude that we are returning to the period of certainty in our behavior. The sooner we come to this awareness, the less systemic losses we will have.

We talked about the certainty of laws in development when evaluating physical laws. Biological laws imply continuous search in different directions, or ambiguity. But this does not mean that humanity will rise to the solid path of development; on the contrary, having defined the principles of relationships with the primordial nature, we will seek new ways of our development in favor of all nature.

We will learn to plan and maintain the whole world in dialectical equilibrium. For the planning of new evolutionary matters, we will need deep and extensive knowledge of the holistic arrangement of the planet; for example, we must learn to control the weather. We must master the complex of subtle technologies for scientifically based actions in the global world. All our actions must take place in harmony with all nature. We need to design the optimal and dynamic construction of society, build new types of ecological settlements, etc.

We have unwittingly begun answering the eternal questions of philosophy: the meaning of life and the destiny of each person in this world. We are the virtual consciousness of nature and the last highest level of the world's complexity.

It turned out that people cannot be assessed and changed in isolation from the evolution of nature. They are both a goal and a means of general processes of self-complication of nature. In the play of nature, people were its developing theme and sense, its actor; then they became co-directors and a critical (self-critical) consciousness.

For a possible transition to a conscious stage of evolution, we need a systematic and comprehensive knowledge of nature. Therefore, I started the book with a review of the fundamental principles of science and philosophy. At the same time, this book cannot be regarded as the absolute reference. It should be considered only as the beginning and the first step in creating a new paradigm of development in the 21st century.

It should be taken into account that the goals set in the book can be achieved only with the common desire of world scholars for integration to solve the urgent problems of our era.

We need to create **a world academy of sciences** earlier than the world government. It should lay the foundations of the world government. First of all, we will create an active scientific structure that should bring together all world trends in science and philosophy. We must create (together) public consciousness with a vision of the common tasks of world evolution.

Bibliography

#	Author	Title	Publisher	Year
1	A.D. Altshul	Hydraulics and Aerodynamics	Moskva Stroyizdat	1987
2	Aristotle	Metaphysics	Izdatelstvo E	2016
3	A.I. Akhiezer	General Physics	Naukova Dumka	1981
4	Christopher Belshaw	12 Modern Philosophers	AST Moskva	2014
5	Henri Bergson	Creative Evolution	Akademichesky Proekt	2015
6	Zbigniew Brzezinski	The Grand Chessboard	Mezhdunarodnye Otnosheniya	1999
7	Jean Baudrillard	The Matrix of the Apocalypse. The Last Fall of Europe	Moskva Algoritm	2015
8	Jean Baudrillard	Simulacra and Simulation	Postum	2015
9	Bill Bryson	A Short History of Nearly Everything	AST Moskva	2017
10	Georg Wilhelm Friedrich Hegel	The Encyclopaedia of the Philosophical Sciences	Mysl	1974
11	Karl Gerth	As China Goes, So Goes the World	Alpina	2011
12	Yuliya Gippenreyter	Introduction to General Psychology	AST	2017
13	Fyodor Girenok	Shapes and Folds	Akademichesky Proekt	2014
14	A.Kh. Gorfunkel	Philosophy of the Renaissance	Vysshaya Shkola	1980
15	John Gribbin	In Search of Schroedinger's Cat	Ripol classic	2017
16	David G. Green	Reinventing Civil Society: Rediscovery of Welfare without Politics	Novoe Izdatelstvo	2009
17	G.V. Grinenko	History of Philosophy	Urait	2014
18	Lev Gumilyov	The Hsiung-nu	Airis Press	2014
19	A.A. Guseynova	History of Ethical Teachings	Triksta	2015
20	Richard Dawkins	The Blind Watchmaker: Why the Evidence of Evolution Reveals a Universe without Design	AST Moskva	2017
21	A.S. Enokhovich	Handbook of Physics and Engineering	Prosveshchenie	1989
22	Kallistrat Zhakov	Logics: The Art of Dispute	Amfora	2015
23	Albert Camus	West: Conscience or Void?	Algoritm	2014
24	Elias Canetti	Crowds and Power	AST Moskva	2014
25	Immanuel Kant	Critique of Pure Reason	Moskva	2015
26	Fritjof Capra	The Tao of Physics	Mann, Ivanov and Ferber	2017
27	André Comte-Sponville	Philosophical Dictionary	Aeterna	2012
28	Joel Kramer	The Passionate Mind Revisited	Dekom	2011
29	Lawrence M. Krauss	Fear of Physics: A Guide for the Perplexed	Piter	2016
30	Aleksandr Kuvshinov	Tao Te Ching	Profit	2015
31	Søren Kierkegaard	Either/Or	Akademichesky Proekt	2014
32	S.A. Lebedev	Philosophy of Science	Akademichesky Proekt	2008
33	Vladimir Levi	The Art of Being Self	Znanie	1991
34	Mario Livio	Brilliant Blunders: From Darwin to Einstein	AST Moskva	2015
35	Herbert Marcuse	Critical Theory of Society	Astrel	2011
36	Vladimir Mezentsev	Encyclopedia of Miracles	Alma-ata	1990
37	Michio Kaku	Physics of the Impossible: A Scientific Exploration Into the World of Phasers, Force Fields, Teleportation, and Time Travel	Alpina	2017

38	Joseph S. Nye Jr.	The Future of Power	AST Moskva	2014
39	I.S. Narsky	Western European Philosophy	Vysshaya Shkola	1974
40	S.A. Nizhnikov	Philosophy	Infra-m	2014
41	Friedrich Nietzsche	Small Collected Works	Azbuka	2014
42	Plato	Dialogs	Akademichesky Proekt	2015
43	N.V. Popkova	Philosophy of the Technosphere	LKI	2008
44	A.D. Potapov	Ecology	Vysshaya Shkola	2014
45	James Redfield	The Tenth Insight: Holding the Vision	AST Moskva	2008
46	Matt Ridley	The Eolution Of Everything. How New Ideas Emerge	Izdatelstvo E	2017
47	Jonathan Sacks	The Dignity of Difference	Gesharim	2008
48	Jean-Paul Sartre	Search for a Method	Akademichesky Proekt	2008
49	Aleksandr Semyonov	Interesting Philosophy	Palmira	2016
50	Lucius Annaeus Seneca	On Living to Oneself	Mann, Ivanov and Ferber	2015
51	Lauren Slater	Opening Skinner's Box: Great Psychological Experiments of the Twentieth Century	AST Moskva	2007
52	Irving Stone	The Origin	Politicheskaya Literatura	1989
53	A.D. Sukhanov	Concepts of Modern Natural Science	Drofa	2004
54	Nassim Nicholas Taleb	On Secrets of Sustainability	Kolibri	2012
55	Arnold Joseph Toynbee	Civilization on Trial	AST Moskva	2008
56	Alvin Toffler	Revolutionary Wealth	AST Moskva	2008
57	T.I. Trofimova	Physics Course	Vysshaya Shkola	2004
58	David A. Wilson	The History of the Future	AST Moskva	2007
59	Herbert George Wells	A Short History of the World	AST Moskva	2011
60	Richard Phillips Feynman	The Character of Physical Law	AST Moskva	2017
61	Sigmund Freud	Totem and Taboo: Resemblances Between the Mental Lives of Savages and Neurotics	AST Moskva	2012
62	Thomas Loren Friedman	Hot, Flat, and Crowded: Why We Need a Green Revolution—And How It Can Renew America	AST Moskva	2011
63	I.T. Frolov	Philosophical Dictionary	Respublika Sovremennik	2009
64	Erich Fromm	To Have or to Be?	AST Moskva	2018
65	Erich Fromm	Freudian Theory	AST Moskva	2012
66	Erich Fromm	The Soul of Man	AST Moskva	2014
67	Francis Fukuyama	America at the Crossroads: Democracy, Power, and the Neoconservative Legacy	AST Moskva	2008
68	Martin Heidegger	Being and Time	Akademichesky Proekt	2013
69	S.D. Haitun	The Phenomenon of Man against the Backdrop of Universal Evolution	Librokom	2009
70	Khara-Dawan Erejen	Genghis Khan as a Commander	Progress	1992
71	Robert Hazen	The Story of Earth. The First 4.5 Billion Years, from Stardust to Living Planet	Alpina	2017
72	S.P. Khromov	Meteorology and Climatology	Nauka Moskva	2006
73	Albert Schweitzer	Reverence for Life	Progress	1992
74	Arthur Schopenhauer	Aphorisms on the Wisdom of Life	Eksmo	2015
75	Oswald Spengler	The Decline of the West	Poppuri	2008
76	Friedrich Engels	Anti-Duhring	Politizdat	1978
77	B.M. Yavorsky	Reference Guide to Physics	Nauka	1984
78	Karl Jaspers	The Spiritual Situation of the Age	AST Moskva	2013

Figures

(Option A) **Kinematics of the planet's motion**

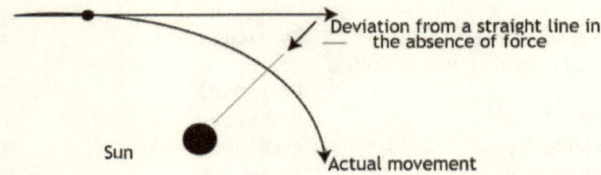

(Option B) **The rule of force parallelogram**

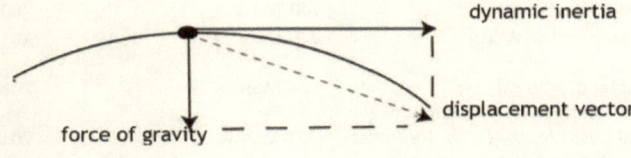

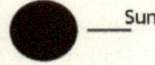

Fig. 1. Kinematics of the planet's motion

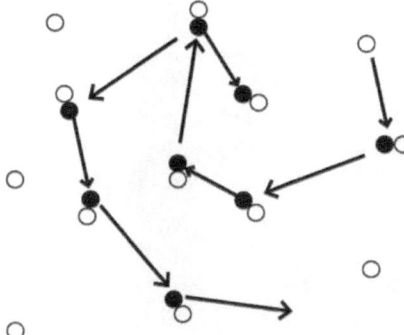
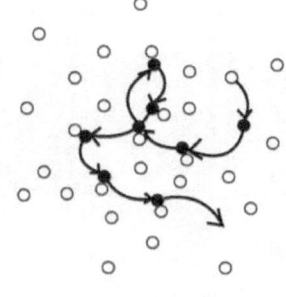

Fig. 2. Trajectory of a gas molecule's motion

Fig. 3. The trajectory of motion of an elementary particle with increasing density of the medium

Clarifying the notion of a discrete medium

The properties of a certain number of gases are determined by the motion of a huge array of its molecules. Any separate molecule moves along the broken line, as shown in the Figure. This set of molecules can be considered as a discrete medium, where each particle has the energy of motion, while there is no interaction force between the particles (the force of attraction is weak).

The accumulation of elementary particles in space (Fig. 6) at the initial stage is a discrete medium. Continued accumulation of amions in space leads to an increase in the density of particles. As the density increases, the accumulation of amions becomes a non-discrete medium (see Fig. 3).

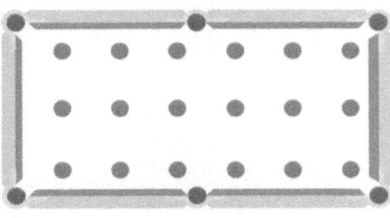

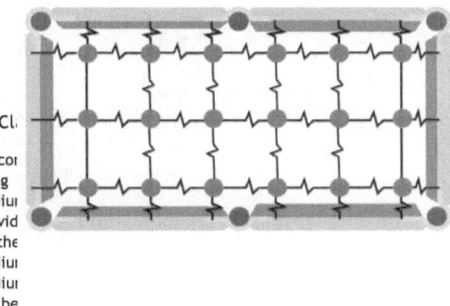

Cl:

In co along mediur individ by the mediur mediur will be

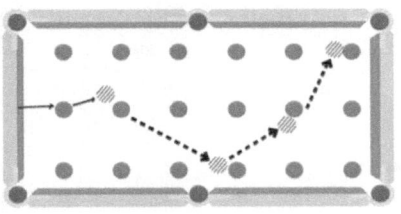

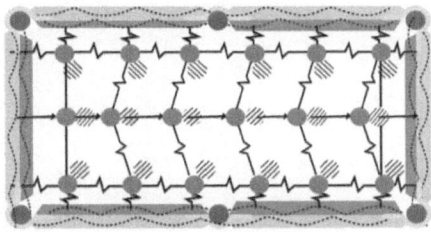

Fig. 5. Amions on the billiard table

Fig. 4. Billiard table

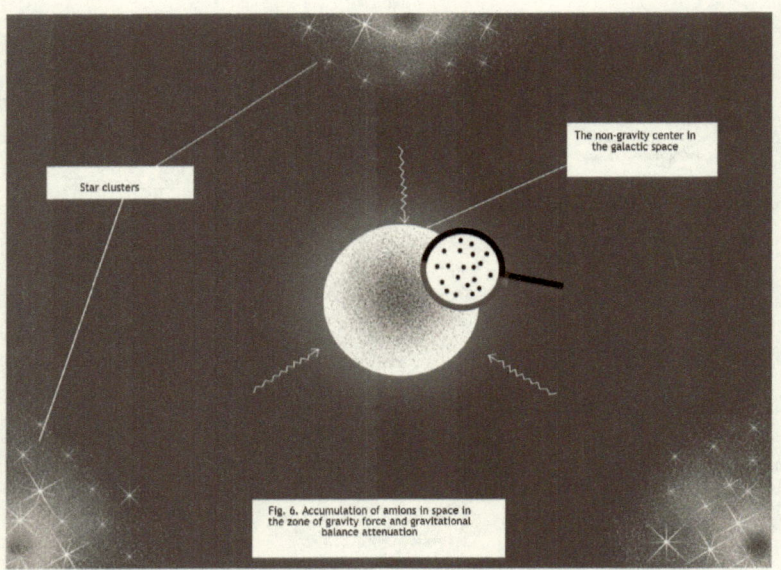

Fig. 6. Accumulation of amions in space in the zone of gravity force and gravitational balance attenuation

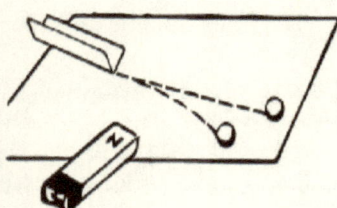

Fig.7. Curvature of the ball's trajectory as a result of interaction with a magnet

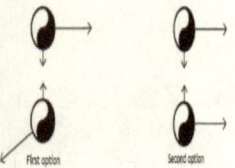

First option

Second option

Fig. 8.

As outer compaction and self-compaction of the aggregate of primary matter in the galactic space occur, infinite interaction (percussion) of the particles occurs between themselves.

This figure shows two options of particle interaction.
Elementary particles move in space by inertia, while particles are attracted to each other by the aseric properties. Only in the second option, if the direction coincides, the particles can approach each other. When a certain level of self-compaction of the medium is reached, the accumulation of particles in space initiates the vortex of the cosmic sphere (see Fig. 10). An infinite motion of elementary particles in a circle is formed.

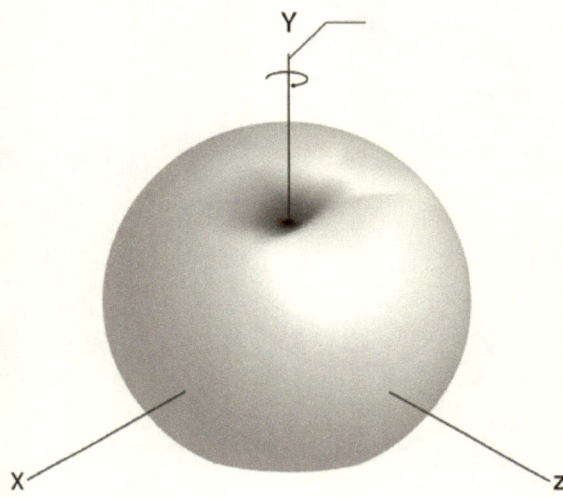

Note. Axis of rotation coincides with the Y axis.

Fig. 9. General view of the cosmic sphere in the coordinate system

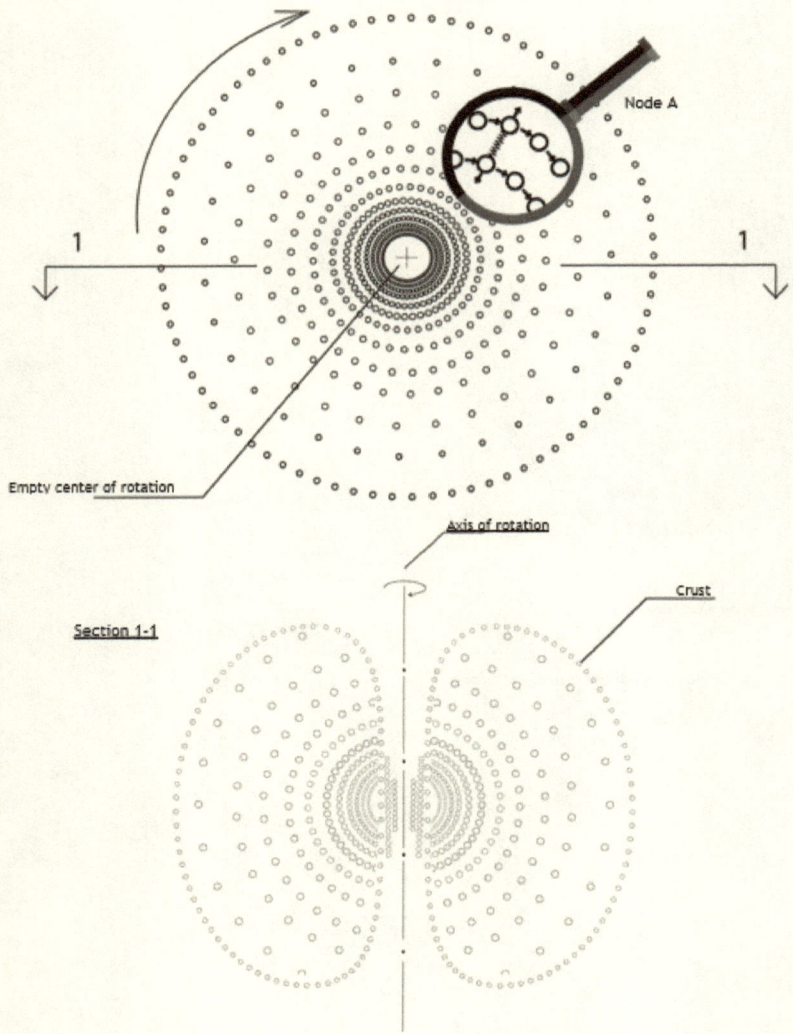

Fig. 10. An arbitrary vortex of the primary matter and the formation of the cosmic sphere

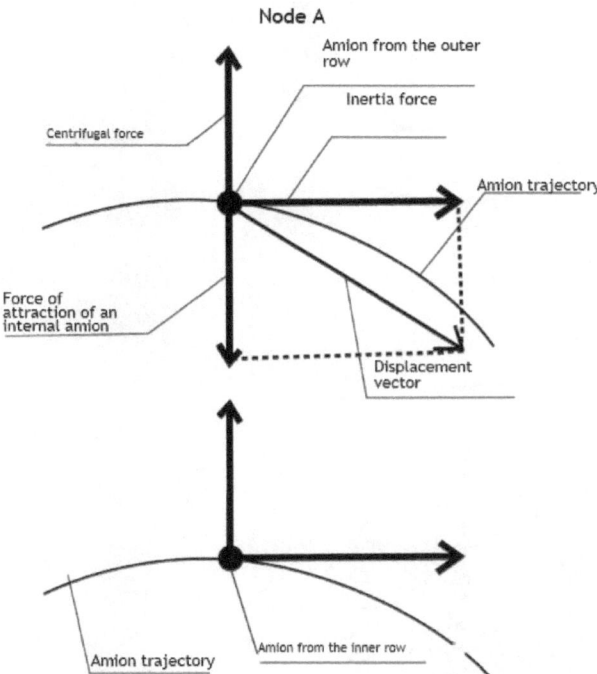

Fig. 11. The appearance of centrifugal force

The material particle (amion) on the ball's periphery tends to move with inertia forces rectilinearly, and the force of attraction of the internal amion changes the particle's trajectory. When the inertial state of the amion changes (during rotation), centrifugal forces appear. Centrifugal forces prevent the compaction of the cosmic sphere in the radial direction.

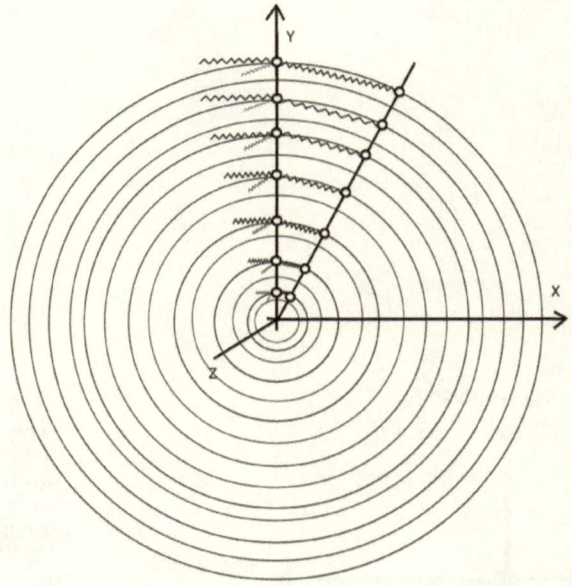

X,Y- ecliptic plane
Z - axis of rotation of the Solar System

Fig. 12. Scheme of three aseric bonds

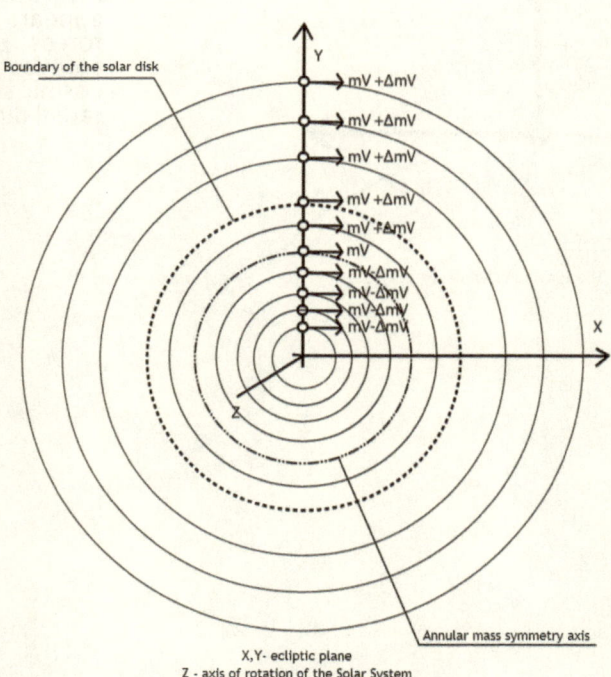

X,Y- ecliptic plane
Z - axis of rotation of the Solar System

Fig. 13. Scheme of redistribution of the inertia norm of amions
along the Y axis (in the radial row)

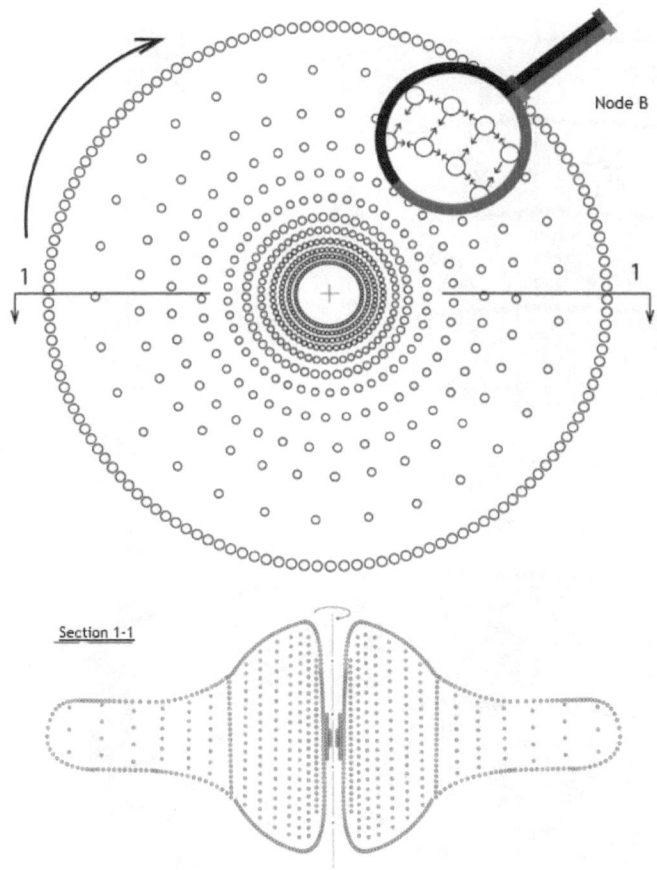

Further compaction of the ellipsoid forms a cosmic 'fried egg.' The compaction of the system's central part is prevented by the factor of limiting compaction of the primary matter.

We will consider the dynamic system (structure) of elementary particles as a semisolid disk. The particles are bound by forces of attraction; at the same time, the bonds can be broken; the medium is like a liquid.

Fig.14. Deformation of the cosmic sphere

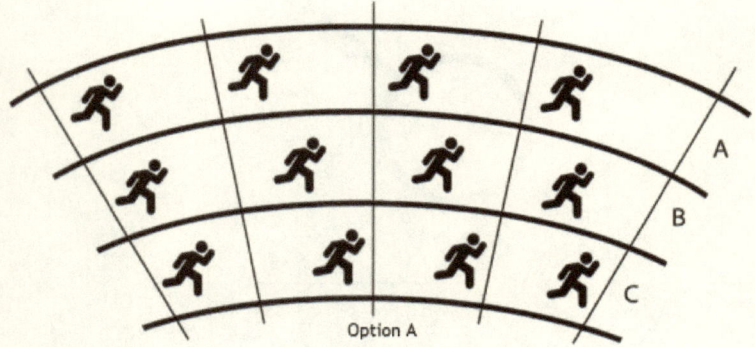

Runners in the outer rows lag behind.

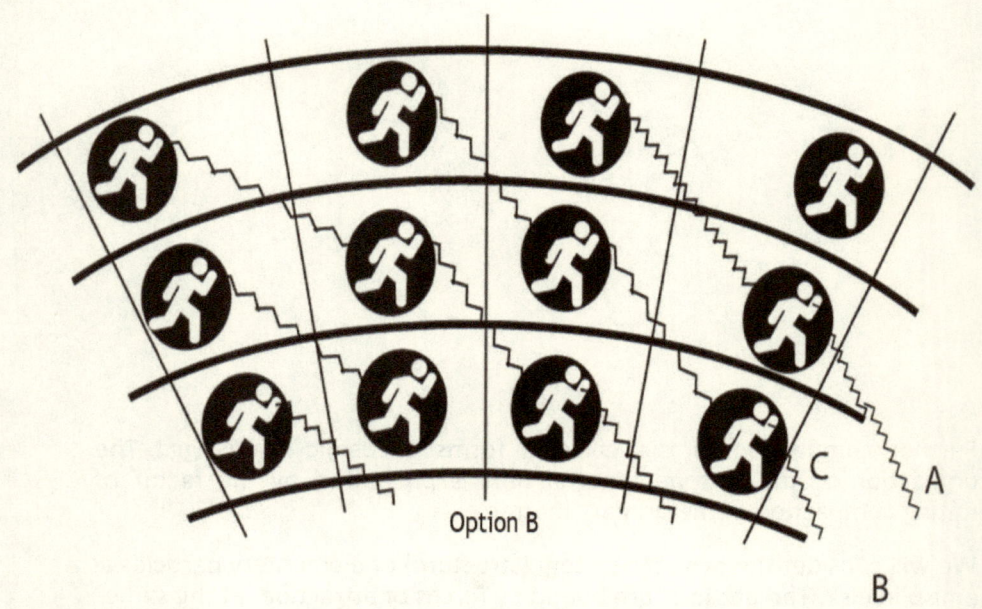

Unlike runners in the stadium, amions along the circular trajectory of the Solar System 'run' in a bunch.

Fig. 15. Movement of runners in a circle

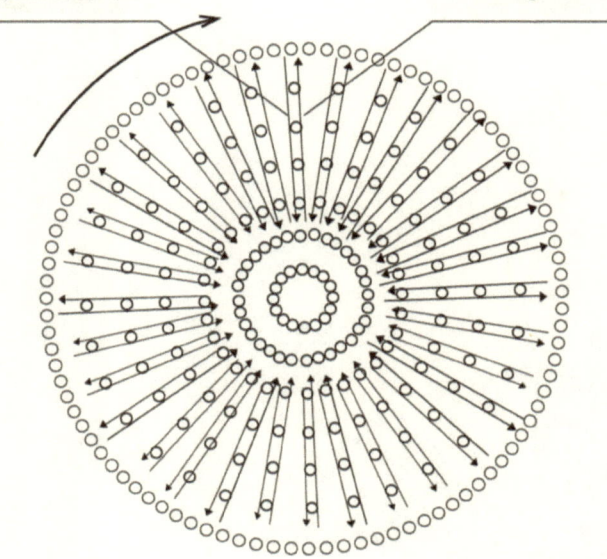

1) Radius vector refers to centrifugal forces; they are created by the amions' tendency to move rectilinearly by inertia. The force of attraction of the amions from the center of the vortex induce (force) the series of amions in the outer rows to move in circles, and this causes the reactive centrifugal forces, which is expressed by the radius vector.
2) At the same time, the radius vector refers to centripetal force. A denser center organizes a string of amions connected by the aseric forces, which compact the entire cosmic sphere.
3) At the initial stage, the centripetal force prevails over the centrifugal force, therefore the sphere is self-compacted, $F>F'$.

This figure shows opposing forces – centripetal against centrifugal. This struggle will subsequently lead to the appearance of the ecliptic. The Solar System will develop in one plane. Later, even complex DNA molecules will have an ecliptic nature.

Fig. 16. Space system – a semi-rigid disk

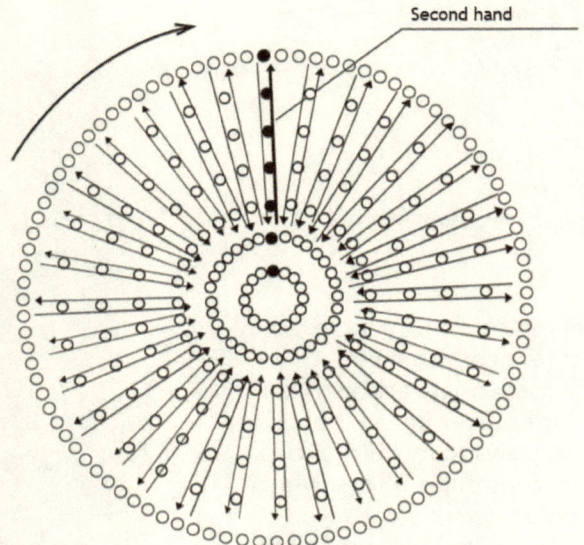

Fig. 16a. Space system - a semi-rigid disk

4)   To show the self-compaction of the cosmic sphere, we introduce the conditional launch situation. In the starting situation, the amions in the vector radius (arrows) zone are marked with black marks.
5)   Further, this arrow will serve as a conventional second hand.
6)   The movement of the sphere occurs by rotating all the amions in a circle.

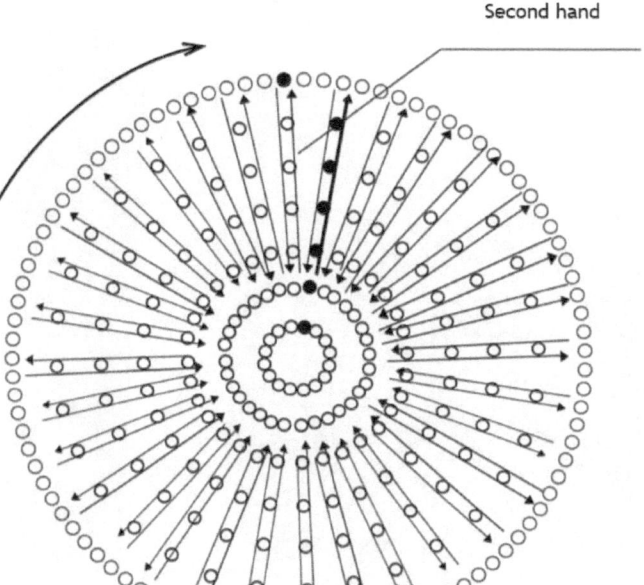

Fig. 16b. Space system - a semi-rigid disk

7) In this group of figures, I show how the big sphere rotates in time. Contradictions immediately occur in the system; amions of outer rows begin to lag behind.

8) While the sphere is very large, the contradictions are solved by a simple breaking of the aseric bonds between the amions in the radial ray (radius vector).

9) The figures show that the marked (black) amions begin to lag behind, breaking the bonds of the centripetal force, and end up outside the vector radius. They are replaced by other, unmarked amions. While the sphere is large and still loose, the contradictions are solved by a simple breaking of the aseric bonds in the radial rows.

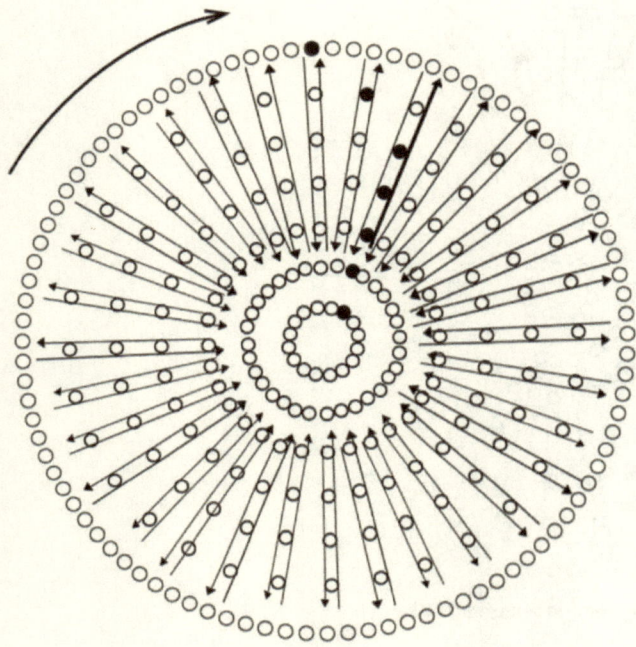

Fig. 16c. Space system - a semi-rigid disk

While the sphere is large, its rotation occurs purely by the force of *energy* (by inertia). The contradictions in the system are solved by breaking the aseric bonds between the amions in the radial row. But in the future, the strengthening of the centripetal force will lead to the compaction of the sphere, which, in turn, will lead to the appearance of the Coriolis force. As a result, a secondary vortex of the amions will begin in the system.

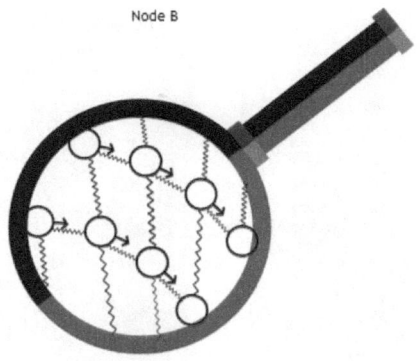

Forces acting on a particle

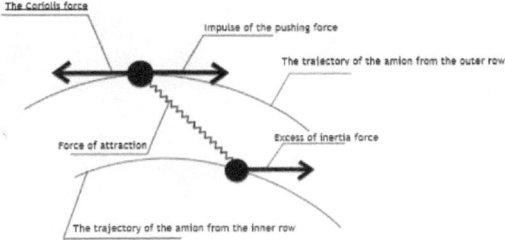

Fig. 17. The appearance of the Coriolis force

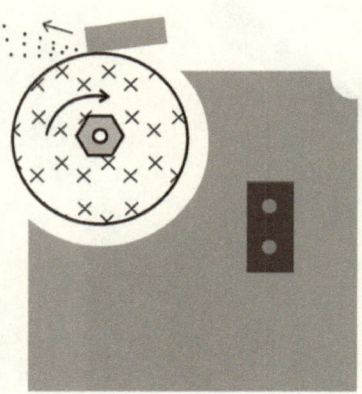

The trajectory of the sparks from the grindstone is directed tangentially to the disk trajectory.
The sparks trajectory is explained by the appearance of Coriolis force.

Fig. 18. An example with a grindstone

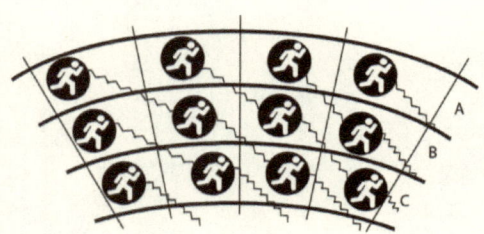

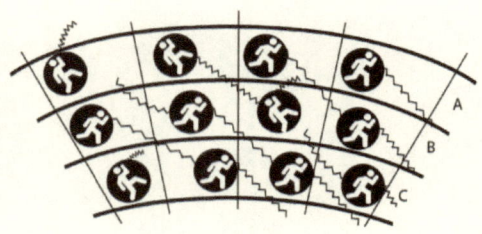

Fig. 19. Breaking of aseric bonds

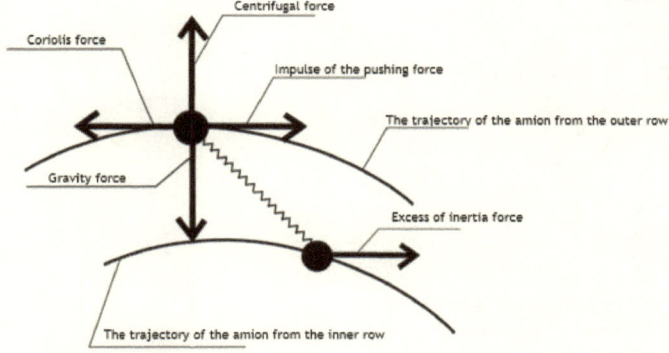
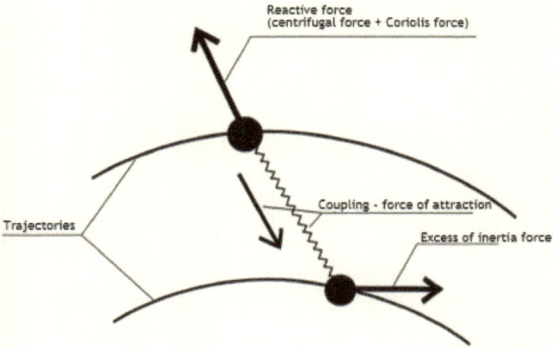

Fig. 20. The occurrence of reactive force

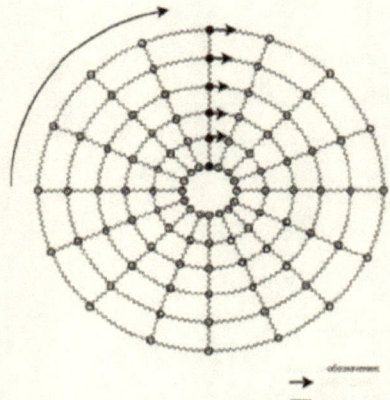

The conditional starting position of amions in
the schematic diagram of the atom.

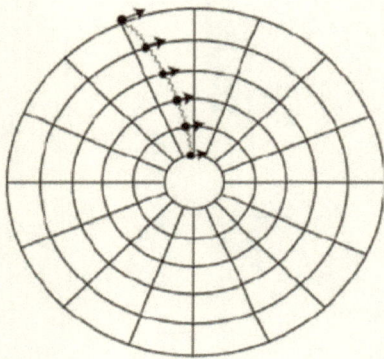

Dissonance in circular motion.
Angular lag of the amines in the outer rows.

Fig. 21. The Stick Effect

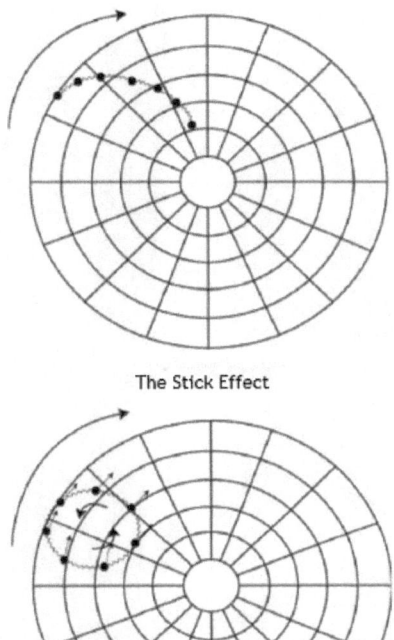

The Stick Effect

Intermediate vortex stage.

Fig. 22. Formation of the internal vortex

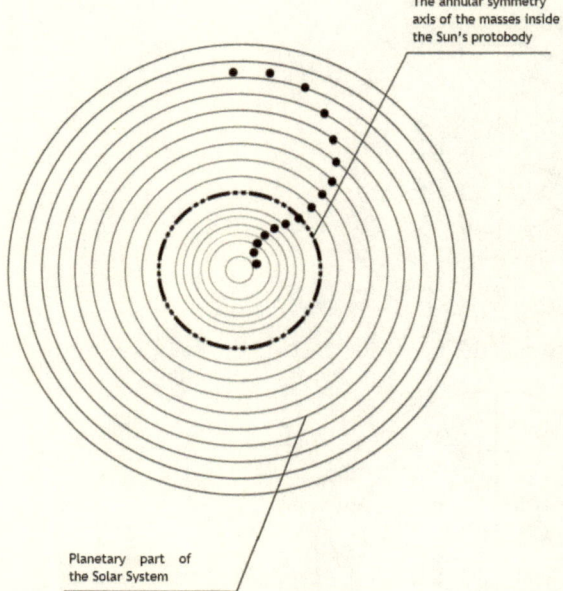

Fig. 22a. The Double Stick Effect

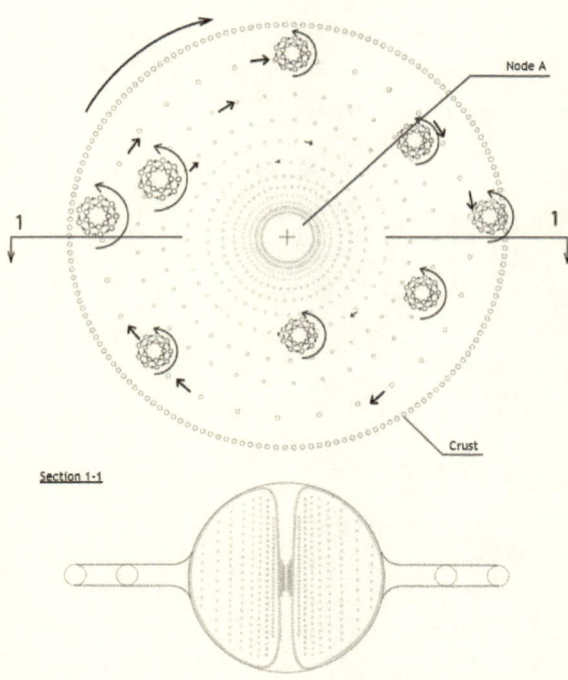

Local vortex occurs on the space system's periphery under the influence of the Coriolis force.

Fig. 23. The beginning of the planetary system formation

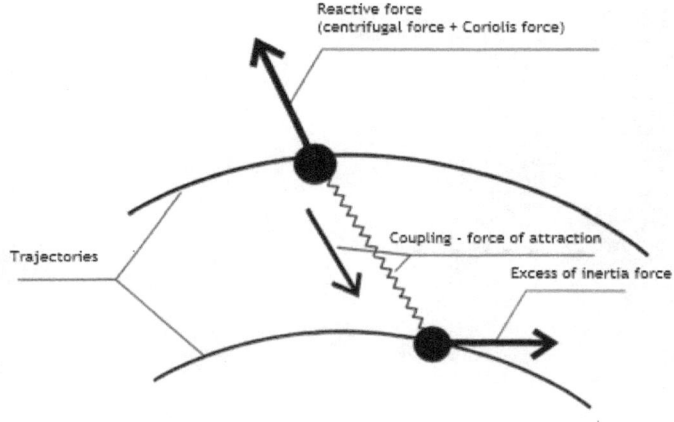
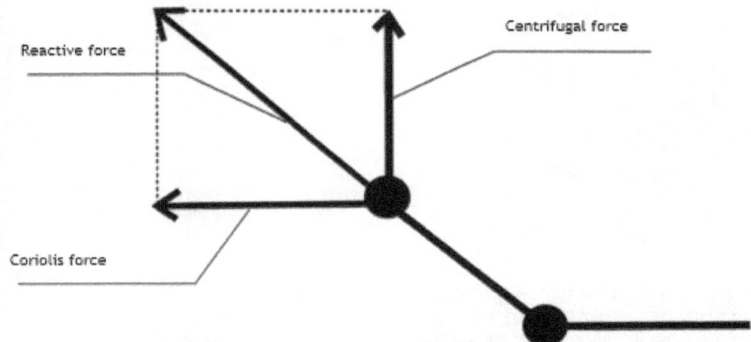

Fig. 24. Decomposition of reactive force

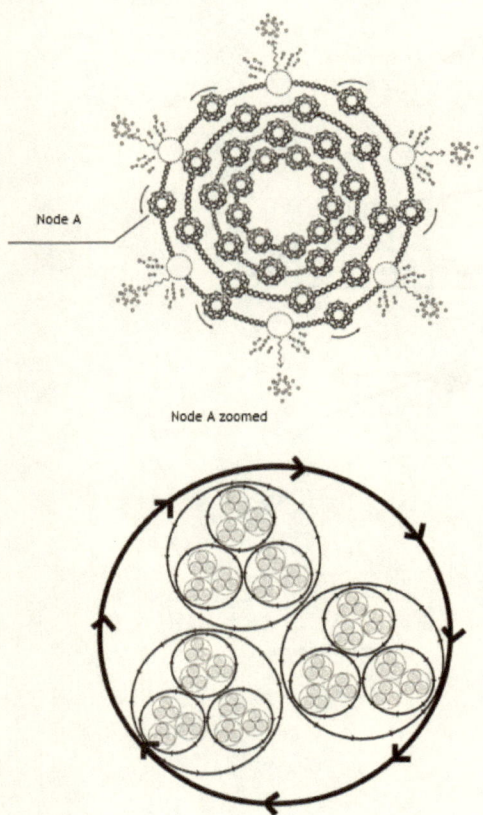

Note. The appearance of three internal vortexes in Node A is conventionally accepted; instead of three, hundreds, thousands, or millions of vortexes may appear. We do not know their exact number; we know only the principle of their appearance. There can be many vortexes; inside these vortexes, a multitude of internal vortexes may appear. Implosion of vortexes can continue until the appearance of nanovortexes.

Fig. 25. Implosive nature of the appearance of vortexes

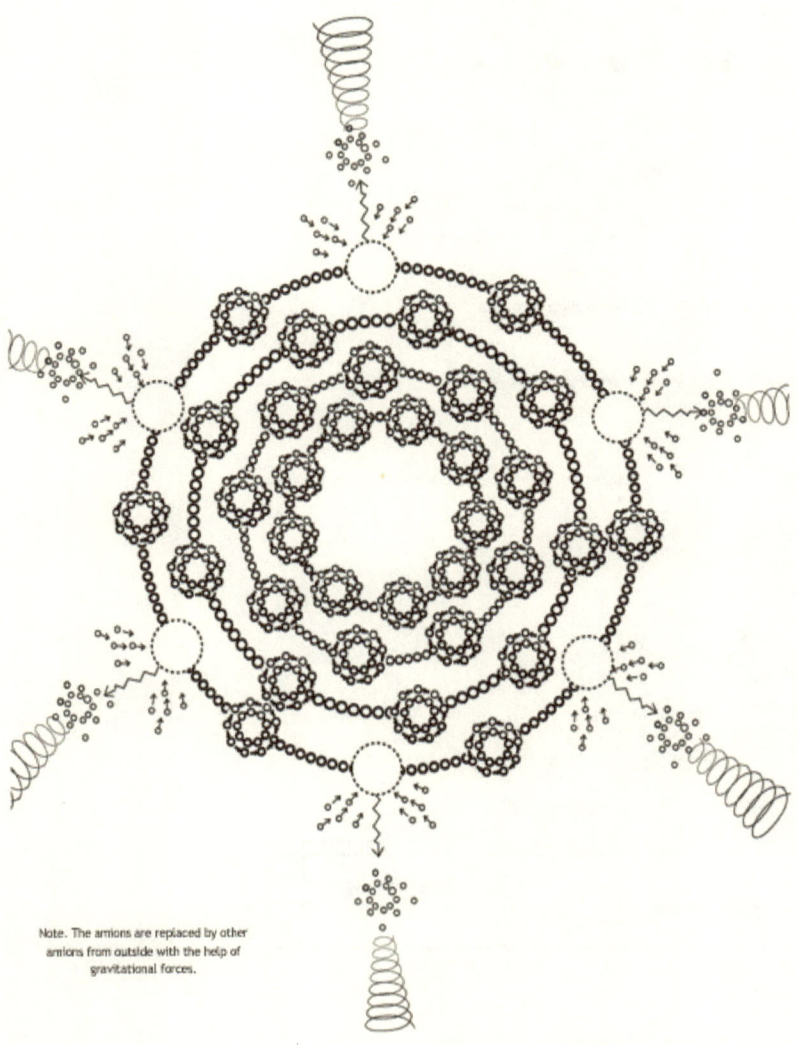

Note. The amions are replaced by other amions from outside with the help of gravitational forces.

Fig. 26. Continuous processes of vortex formation in the solar disk is the cause of solar radiation

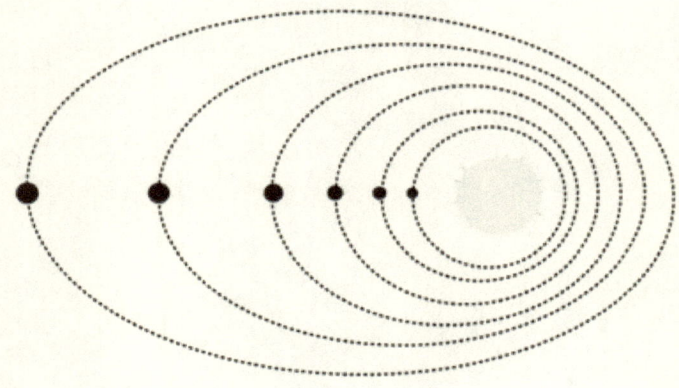

Fig. 27. Stretching of orbits

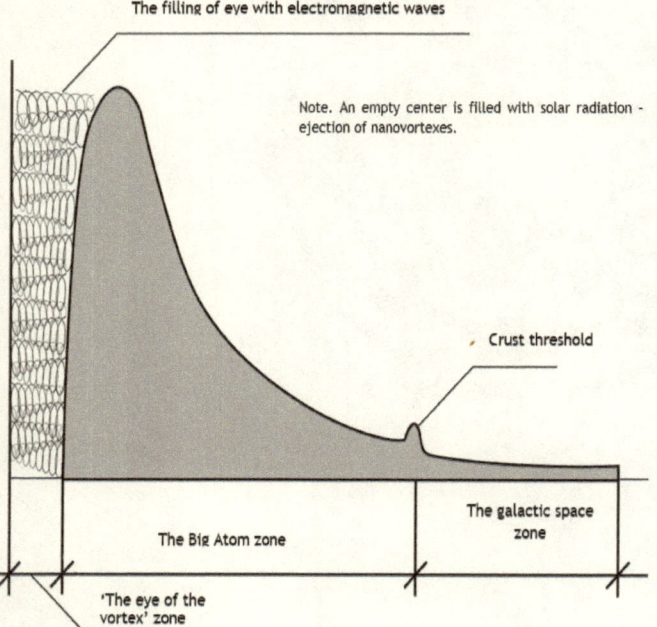

Fig. 28. Entropy of the Big Atom

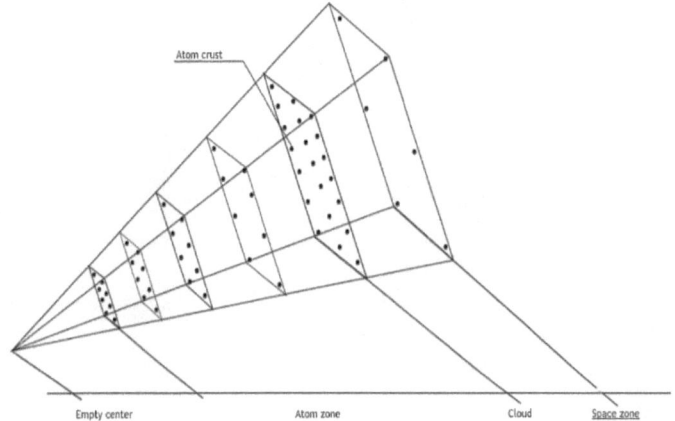

This figure shows four sections of the microscopic segment of the atom. In all sections, the number of amions is the same; in the last, fifth section, in the region of the cortex, the number of amions is larger. Outside the atom, in the atom cloud, the number of amions decreases sharply. Outside the crust appears a cloud, which is a continuation of the atom in the amionosphere. The continuation of the cone in the amionosphere will be of very low density. The cloud density will decrease in the quadratic function.

Fig. 29. Spatial segment of the atom

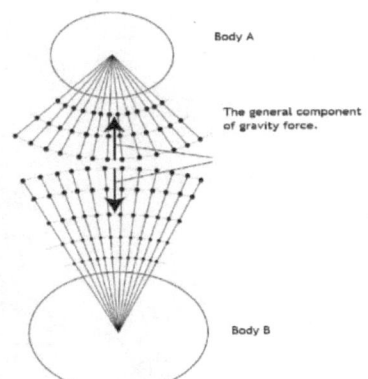

The meeting of amionospheres of two bodies.

Fig. 30. Interaction of gravity force of bodies A and B

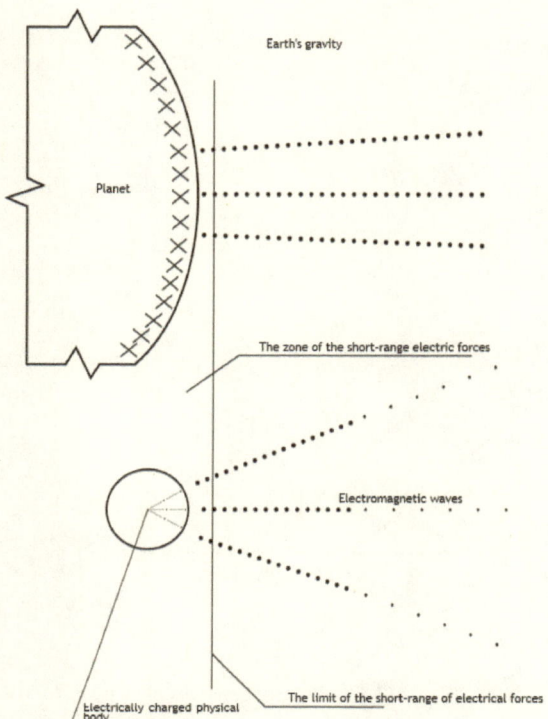

Fig. 31. The nature of short-range action

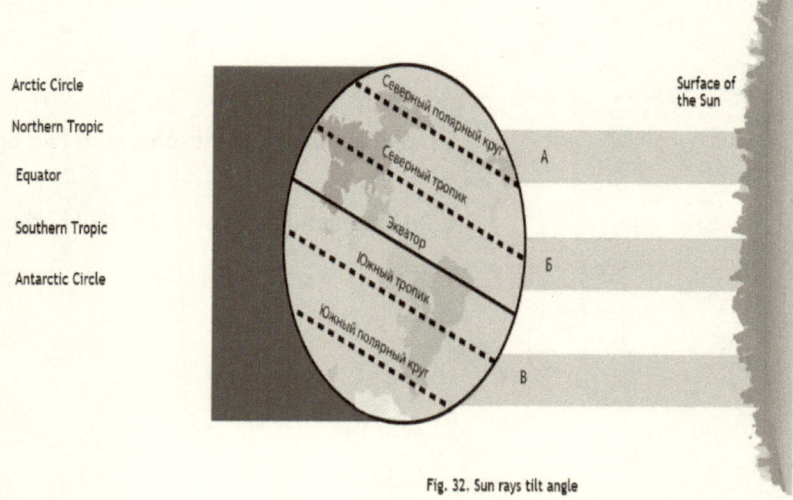

Fig. 32. Sun rays tilt angle

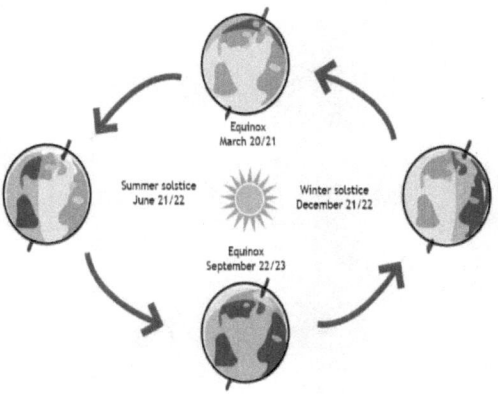

Fig. 33. Annual cycle of the planet

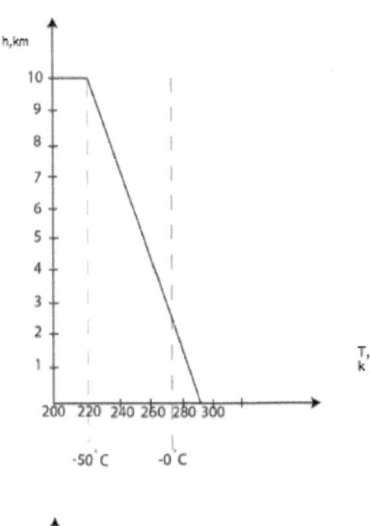

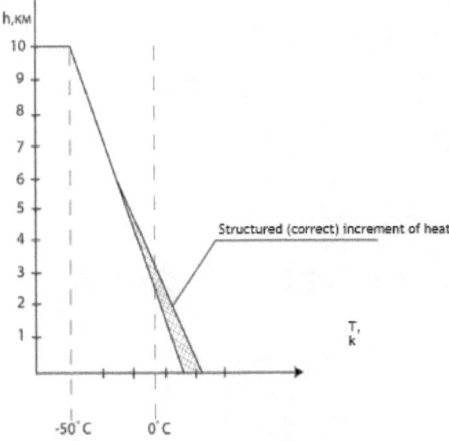

Fig. 34. Temperature of the atmosphere at different altitudes above the ground

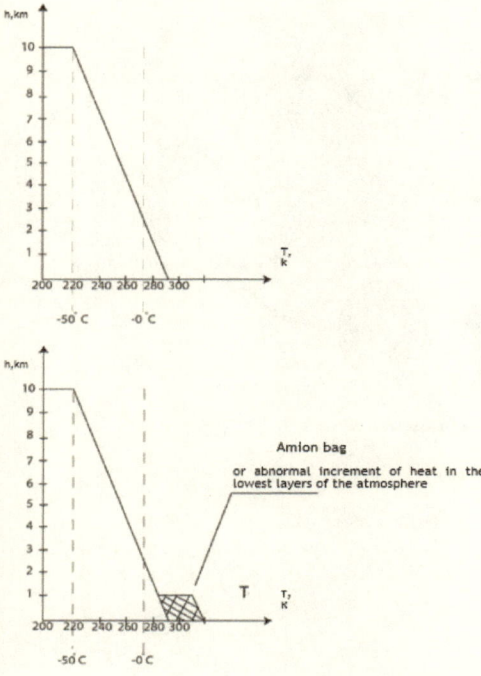

Fig. 35. Temperature of the atmosphere at different altitudes above the ground

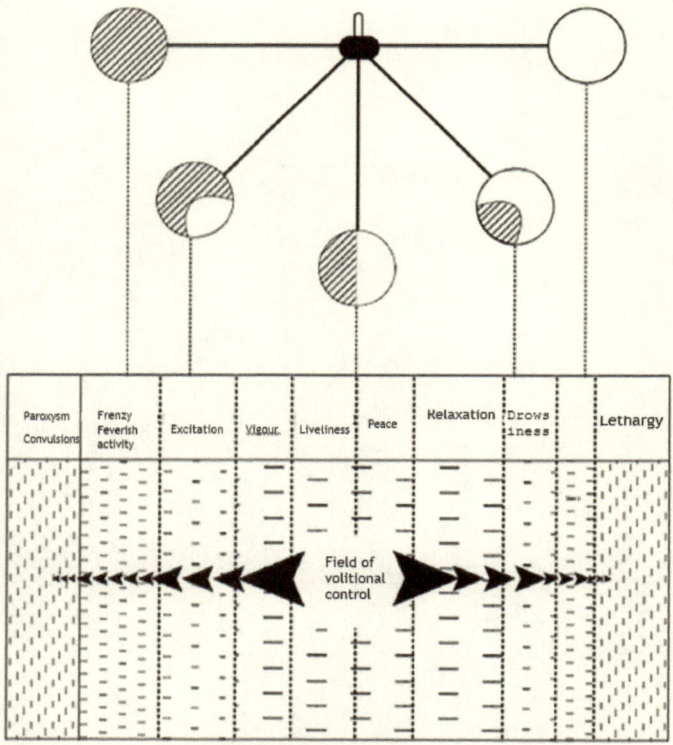

Fig. 36. The pendulum of activity

Periodic table of evolutionary transformations

Stage 1. Structuring of the Solar System

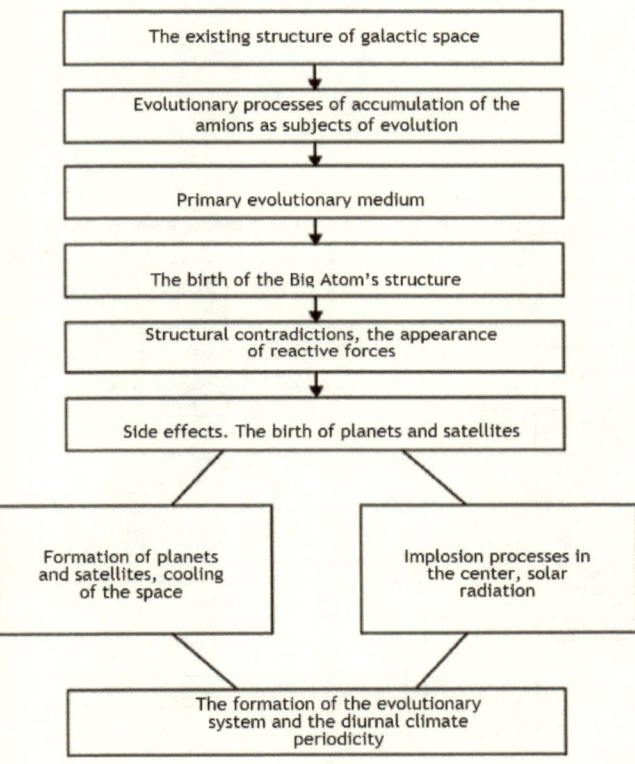

Fig. 37

Periodic table of evolutionary transformations (continued)

Stage 2. Hot stage of evolution. Planet structuring

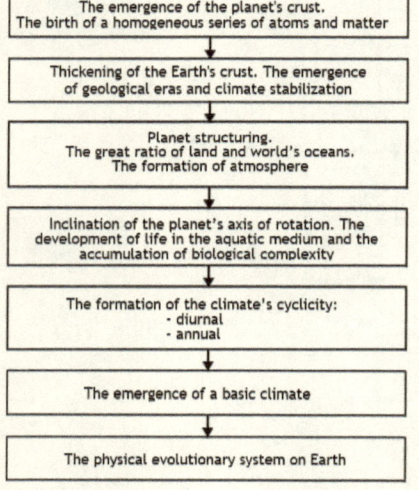

Periodic table of evolutionary transformations (continued)

Stage 3. The slowdown of the physical evolution of the world and the onset of the biological evolution on land.

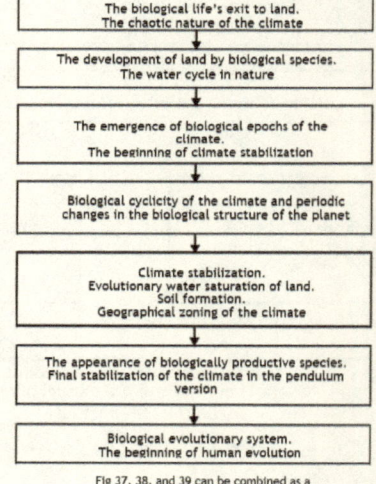

Fig. 38

Fig. 39

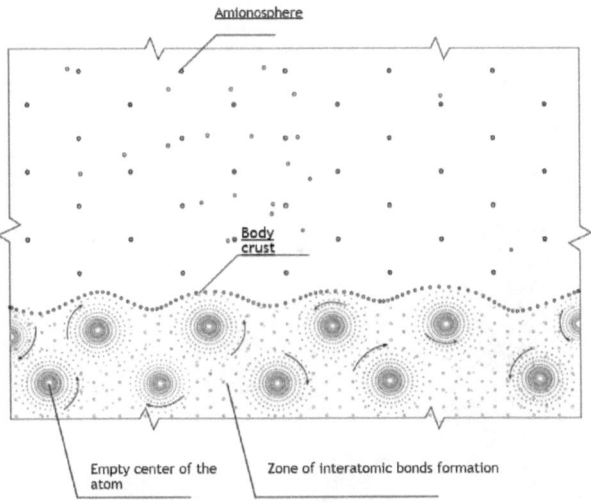

Fig. 40. Structure of the substance protobody

Let's reduce the Earth thousandfold, and increase the amions billionfold.

Imagine the amions in the form of a string of rubber balls attached to the Earth.

In reality, the aseric bonds or properties of attraction will serve as the strings.

In the figure, we can see that the amions have more freedom to move horizontally.

Fig. 41. The nature of gravity forces

This figure also shows the horizontal aseric bonds. Together, radial and horizontal bonds make up the Earth's volumetric gravity network or its amionosphere. As the distance from the Earth grows, the density of the gravitational network decreases in the quadratic function. This explains the inverse dependence of the force of universal gravitation on the distance. Having more freedom of movement in the horizontal plane, the amions can move only as a common array. In certain cases, 'unemployed' amions can appear in the gravitational system (amionosphere), which can stimulate the vortex formation in the medium.

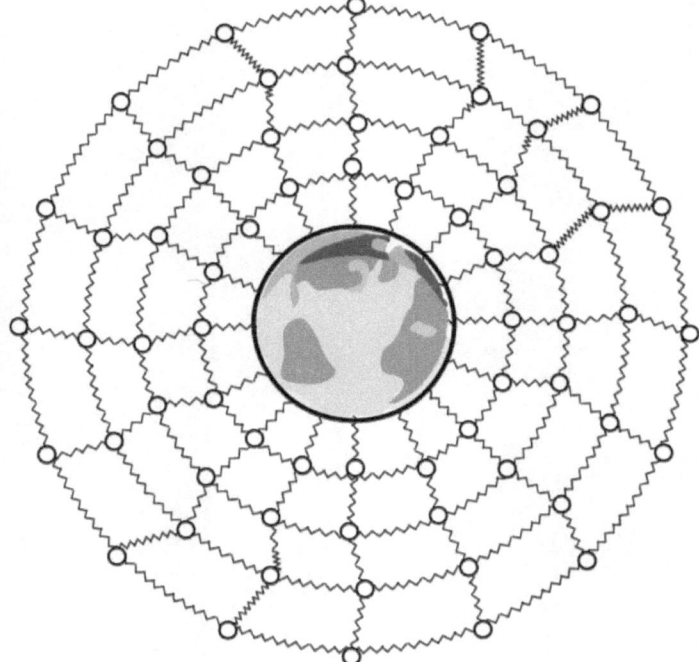

Fig. 42. The gravitational network of the Earth. Amionosphere

www.ingramcontent.com/pod-product-compliance
Lightning Source LLC
Chambersburg PA
CBHW031632210526
45464CB00004B/1859